BIOPESTICIDES MANUAL

Guidelines for Selecting, Sourcing, Producing and using Biopesticides for Key Pests of Tobacco

BIOPESTICIDES MANUAL

Guidelines for Selecting, Sourcing, Producing and using Biopesticides for Key Pests of Tobacco

Keith Holmes, Dirk Babendreier, Melanie Bateman,
Malvika Chaudhary, Julien Grunder, Margaret Mulaa,
Léna Durocher-Granger, Muhammad Faheem

CABI is a trading name of CAB International

CABI	CABI
Nosworthy Way	745 Atlantic Avenue
Wallingford	8th Floor
Oxfordshire OX10 8DE	Boston, MA 02111
UK	USA
Tel: +44 (0)1491 832111	Tel: +1 (617)682-9015
Fax: +44 (0)1491 833508	E-mail: cabi-nao@cabi.org
E-mail: info@cabi.org	
Website: www.cabi.org	

A catalogue record for this book is available from the British Library, London, UK.

ISBN-13: 978 1 78924 202 7   (paperback)
          978 1 78924 201 0   (eBook)

Commissioning Editor: Rebecca Stubbs
Editorial Assistant: Emma McCann
Production Editor: Tim Kapp

Typeset by Exeter Premedia Services Pvt Ltd, Chennai, India
Printed and bound in the UK by Severn, Gloucester

Table of Contents

List of Figures

List of Tables

List of Acronyms

AI	active ingredient
ALP	Agricultural Labor Practices
ANOVA	analysis of variance
AS	active substance
BLM	basal liquid medium
Bt	*Bacillus thuringiensis*
BYSB	brewer's yeast/sucrose broth
cfu	colony-forming units
CORESTA	Cooperation Centre for Scientific Research Relative to Tobacco
CPA	crop protection agent
CRD	Completely Randomized Design
EPN	entomopathogenic nematodes
EPPO	European and Mediterranean Plant Protection Organization
ETRF	experimental *Trichogramma* rearing facility
EU	European Union
FAO	Food and Agriculture Organization
GAP	good agriculture practices
HHP	highly hazardous pesticide
IPM	integrated pest management
MRL	maximum residue level
MYE	molasses yeast extract
NGO	non-governmental organization
NPV	nucleopolyhedrosis virus
OECD	Organization for Economic Co-operation and Development
OV	organic vapour
PCA	potato carrot agar
PDA	potato dextrose agar
PDB	potato dextrose broth
PHI	pre-harvest interval
PIB	polyhedral inclusion bodies
PPE	personal protection equipment
RCBD	Randomized Complete Block Design
REI	re-entry interval
r.h.	relative humidity
SADC	Southern African Development Community
SAR	systemic acquired resistance
SDA	Sabouraud dextrose agar
SDW	sterile distilled water
US EPA	United States Environmental Protection Agency
UV	ultraviolet
WHO	World Health Organization

Acknowledgements

This e-book has been made possible through funding and support from Philip Morris International. CABI retained complete independence and freedom in producing this publication. We also gratefully acknowledge the support of CABI UK, in particular Sarah Thomas, Emma Thompson, Belinda Luke and Steve Edgington.

Disclaimer

Although CABI has taken reasonable care to ensure that the information, data, and other material made available is accurate and up-to-date, CABI accepts no responsibility for any changes to this publication thereafter, including but not limited to any defects caused by the transmission or processing of the information, data and other material. The information made available, including any expression of opinion and any projection or forecast, has been obtained from or is based upon sources believed by CABI to be reliable but is not guaranteed as to accuracy or completeness. The information is supplied without obligation and on the understanding that any person who acts upon it or otherwise changes his/her position in reliance thereon does so entirely at his/her own risk. Information supplied is neither intended nor implied to be a substitute for professional or medical advice. PMI is not responsible for this publication. Please follow this link to view the full terms and conditions: http://www.cabi.org/terms-and-conditions

1 Introduction to the Biopesticides Manual

1.1 Biopesticides, a Key Component of Integrated Pest Management in Tobacco

When used judiciously, crop protection agents (CPAs) such as insecticides, fungicides and herbicides can play an important role in plant protection, reducing the impact of pests[1] on the yield and quality of tobacco. By their nature, CPAs affect living organisms so there are also hazards associated with their use. When CPAs are not used appropriately, they can become a cause for concern due to the risks that they pose to the health and safety of farmers, farm workers and consumers, as well as their potential impact on the environment. Likewise, indiscriminate use of CPAs can exacerbate pest problems through the loss of natural pest control mechanisms and the development of pest resistance. The public in general and consumers in particular are sensitive to these health and environmental concerns. Thus, the risks posed by CPAs must be managed by following strict guidelines for their use.

The implementation of integrated pest management (IPM) can help to address these issues. IPM is a shift in all parts of the supply chain to an ecosystem approach, promoting best practices for the prevention and management of pests in tobacco. IPM seeks to reduce the risk of harm to people and the environment. Attaining the objectives of IPM will depend in particular on changes in farmers' behaviour so that they reduce unnecessary CPA use, use the least hazardous of those CPAs that are registered for use in tobacco, and manage CPAs appropriately.

A key part of an IPM approach is the identification and use of sustainable solutions for managing pests. Biopesticides such as microbials, botanicals, semiochemicals, predators and parasitoids can be an integral tool for IPM strategies, and in some cases they can be a compelling alternative to conventional pesticides. They are often deployed to control insect pests but

---

[1] Any species, strain or biotype of plant, animal or pathogenic agent injurious to plants or plant products [Glossary of phytosanitary terms. International Standards for Phytosanitary Measures No 5. FAO, 1990; revised 2015].

© CAB International 2019. *Biopesticides Manual: Guidelines for Selecting, Sourcing, Producing and using Biopesticides for Key Pests of Tobacco* (K. Holmes *et al.*)

> **Box 1: Overview of biopesticides**
> - **Microbial biopesticides** consist of microorganisms (e.g. bacteria, fungi, viruses, viroids or protozoa) or their products (metabolites, e.g. protein toxins) as the active substance. Entomopathogenic nematodes are sometimes classed as microbial pesticides.
> - **Macrobials** (macroorganisms) include insects' natural enemies (e.g. parasitoids such as *Trichogramma* wasps or predators such as coccinellid beetles) and entomopathogenic nematodes (though the latter are often considered as microbials).
> - **Biochemical biopesticides** are a diverse group that includes naturally derived biochemicals such as **plant extracts/botanicals**, which are derived from plants and are active against the target pest or pathogen. Botanicals may have direct effects on the target pest or indirect effects via the host plant. Biochemical biopesticides may also be based on metabolites derived from fermentation of living microorganisms e.g. Spinosad.
> - **Semiochemicals** are naturally occurring chemicals emitted by plants, animals and other organisms (which may be synthetically produced) that modify insect pest behaviour. These can be used as repellants, attractants for use with traps, or for mating disruption.

may also be used to target other pests such as microbial pathogens, nematodes, weeds and molluscs.

Biopesticides are CPAs that are either derived from living organisms or are the products of living organisms that can be used to manage pests such as insects, diseases and weeds. For the purposes of this manual, we define biopesticides to include microbials (e.g. bacteria, algae, protozoa, viruses and fungi), macrobials (e.g. predatory insects, parasitoids and beneficial nematodes), botanicals, and semiochemicals. For more information, see Box 1.

When used in conjunction with good crop management, biopesticides can help to keep pest levels under control, reducing the need to apply conventional pesticides. Tobacco farmers must produce quality crops according to good agricultural practice (GAP), which do not exceed maximum residue levels (MRL), and this must often be achieved with a diminishing number of active ingredients (AI) in their pesticide portfolio. Biopesticides are a good option for farmers to use so they are able to comply with GAP while staying below MRL.

1.2   About the Biopesticides Manual

While many in the tobacco sector are actively promoting the uptake of biopesticides in tobacco pest management strategies, there are also challenges which, if not addressed, can impede these efforts. Before a biopesticide can be deployed successfully by farmers in the field, several steps must take place:

- Pest management needs must be understood.
- Registered products containing biopesticide active substances that are effective against the target pests have to be identified.
- Biopesticides have to be available, either commercially or through local production.
- Farmers and the field technicians who advise them have to know how to use the biopesticides properly so that they can achieve good results.

Carrying out each of these steps requires the input and engagement of leaf-supplier decision makers, trial managers, field technicians and the farmers themselves. Each of these groups needs access to information in order to carry out the activities for which they are responsible. Unfortunately this information is not always readily available.

This *Biopesticides Manual: Guidelines for selecting, sourcing, producing and using biopesticides for key pests of tobacco* (hereafter referred to as the '*Biopesticides Manual*') aims to make information resources and technical advice available in order to support the deployment of biopesticides. The *Biopesticides Manual* is intended to be a one-stop shop to address the information needs of the key groups who are responsible for selecting, sourcing and using biopesticides in the tobacco production system. These groups represent the target audiences, all of whom have challenges they must address and varying information needs, which are summarized as follows.

- **Selecting biopesticides:** One challenge faced by some leaf suppliers seeking to include biopesticides in their pest management strategies is a lack of information about which active substances target the major pests of their crop of interest, the registration status of products containing these active substances, the availability of products and how the efficacy of products can be assessed. *Chapter 2* provides information for decision makers to support selection of biopesticide active substances. It also provides guidelines for trial managers on experimental design, data collection and reporting.
- **Sourcing biopesticides:** Once suitable biopesticide active substances are identified, the ground teams are not always in the position to implement them due to issues with sourcing the biopesticides in sufficient quantities and quality for the contracted farmers to use. *Chapter 3* provides guidance for sourcing biopesticides. It also includes manuals for the local production of three types of biopesticide: *Trichogramma*; neem; and fungal biopesticides such as *Trichoderma*.
- **Training for field technicians and farmers on how to use biopesticides:** Working with biopesticides will be something new for many farmers and even some field technicians. Many field technicians and farmers are not familiar with their usage. The provision of training and guidelines is essential for the correct use of biopesticide products and for the successful uptake of this technology. *Chapter 4* presents training materials

to provide an overview of biopesticides in general together with detailed information on how to work with the key biopesticides that have already been used successfully to manage key pests in tobacco. The training activities that are provided in **Chapter 4** are participatory in nature.

2 Selecting Biopesticides

2.1 Identifying Needs and Biopesticide Management Options

Topics covered in this section include:

- Motivations for incorporating biopesticides into IPM strategies
- Decision matrix for selecting biopesticides

2.1.1 Motivations for incorporating biopesticides into IPM strategies

The decision to incorporate a biopesticide into an IPM strategy for the management of a pest may be motivated by a number of reasons:

- Dependence on just a few AI raises the risk of the development of pest resistance to those control measures. When used in an IPM programme, biopesticides can be a good tool to help avoid the development of resistance to conventional CPAs. Biopesticides usually work by using multiple modes of action, which means there is a much lower risk of pests developing resistance to them.
- For some pest problems, either the conventional CPAs that are registered may provide inadequate control or there may not be a CPA registered for management of that particular pest at all. When used in conjunction with good crop management, biopesticides can help to keep pest levels under control, reducing the need to apply other CPAs.
- Registering conventional CPAs requires action from industry, and can be both expensive and time consuming. If a market is considered to be too small, manufacturers may not even pursue new registrations. On the other hand, some countries have policies in place that promote the registration and use of biopesticides. For example, the data requirements may be reduced; registration fees may be lower; the registration process may be accelerated or prioritized; there may be governmental support available for trials; and certain products containing certain types of active substances (AS) may not have to be registered.
- Some CPAs tend not to be applied according to the label requirements. For example, farmers may apply them late in the season in contravention of the pre-harvest interval (PHI), leading to residue concerns. For most biopesticides, residues are not an issue and the PHI may even

allow for the product to be applied on the same day as harvest. Many biopesticides are explicitly exempted from the requirement of an MRL.

- Many of the CPAs registered for use in tobacco are problematic because they meet one or more of the criteria for designation as highly hazardous pesticides (HHPs). The international community recognizes HHPs as a significant issue and has called for concerted action to address them, including their progressive banning (FAO, 2016). As an alternative to the use of hazardous CPAs, biopesticides can be considered as they are generally classed as a categorically 'lower risk' pest management option. On average, their toxicity profiles tend to be much lower than the average conventional CPA (although there are many low toxicity synthetic CPAs). With few exceptions, most biopesticides do not meet any of the criteria for classification as HHPs.

2.1.2   Decision matrix for selecting biopesticides

Biopesticides can potentially be incorporated into IPM strategies to address the needs outlined in Section 2.1.1. However, not all biopesticides will be appropriate, so certain key points must be taken into consideration when selecting biopesticides for use.

To help assess whether or not a particular biopesticide would be an appropriate choice, a decision matrix is provided (Table 2.1). This decision matrix can be applied to all CPAs, not just biopesticides. It covers many of the points that regulators consider when deciding whether or not to register a CPA (FAO, 2017).

2.2   Availability

Topics covered in this section include:

- Overview of the regulatory framework for biopesticides in the countries where tobacco is grown
- Determining which biopesticides are allowed for use in a country

2.2.1   Overview of the regulatory framework for biopesticides in the countries where tobacco is grown

Regulatory frameworks for biopesticides differ by country. In fact, even the set of compounds classified as biopesticides changes from country to country. For many countries where tobacco is grown, there is no legal definition of what constitutes a biopesticide. For the countries where national regulations do define what is a biopesticide, there are many variations in the definition applied. While certain substance groups are generally accepted as being 'biopesticides', for certain other substance groups there is no consensus.

Table 2.1 Decision matrix.

Biopesticide:		
Target pest(s):		
Is the biopesticide effective against the target pest(s)?	● Is evidence available that the biopesticide Is effective against the target problem? Is the control measure known to work reliably under normal farm conditions?	● If **yes**, then proceed to the next point. ● If **no** evidence is available that the biopesticide controls the target pest in tobacco, it should be rejected.
Is the biopesticide safe?	● Are the risks posed by the biopesticide to human health and the environment acceptable?	● If **yes**, then proceed to the next point. ● If **no**, ○ Biopesticides that meet any of the HHP criteria are considered to pose an unacceptable hazard and should be rejected. ○ Where there are other serious human health (e.g. endocrine disruption) or environmental hazards (e.g. bioaccumulation, aquatic toxicity) mitigation measures should be put in place to reduce risk. For example, the biopesticide should only be adopted for use if appropriate personal protection equipment (PPE) is available.
Is the biopesticide sustainable?	● Does evidence indicate that the biopesticide will not affect agronomic sustainability? Is the risk of the development of pest resistance low? Does it pose low risk to pollinators, natural enemies and other beneficial organisms?	● If **yes**, then proceed to the next point ● If **no**, ○ Assess whether mitigation measures can be put in place to reduce risk. ○ If not, the biopesticide should be rejected.

Continued

Table 2.1 Continued

Biopesticide:
Target pest(s):

Is the biopesticide practical?	• Given the local circumstances, is the biopesticide practical for farmers to use? Is its use realistic given farmers' time and labour constraints? Are appropriate application equipment and storage facilities available? Is the biopesticide compatible with other crop protection measures applied in the production system? Is the biopesticide appropriate for use by small-scale farmers (or is it only effective when used for area-wide management)?	• If **yes**, then proceed to the next point.
		• If **no**,
		  ○ Assess whether the practicalities can be overcome, e.g. by adjusting the production system.
		  ○ If the impracticalities cannot be overcome the biopesticide should be rejected.

Is the biopesticide locally available?	• Is the biopesticide registered for use in the country?
	• Can the biopesticide be sourced locally?
		• If **yes**, then proceed to the next point.
		• If **no**,
		  ○ For particularly compelling biopesticides with strong evidence of efficacy, consider exploring the possibility of registration.
		  ○ For biopesticides that are registered for use but not locally available, liaise with manufacturers or consider establishing production facilities.

Is the biopesticide economic?	• Will the revenue be greater than the costs associated with the biopesticide? (c.f. 2.4)	• If **yes**, then there are no impediments to using the biopesticide, suggesting that it will be appropriate to incorporate into an IPM strategy. • If **no**, the biopesticide should be rejected.

Examples include plant-incorporated protectants, antibiotics, microbial fermentation products, etc.

In many countries, the regulation of microbial biopesticides, semiochemicals and botanicals generally takes place within the regulatory system that was designed for the regulation of synthetic CPAs.

Most countries do not include macroorganisms in their lists of registered pesticides. With very few exceptions, regulation of the use of macroorganisms does not fall under the mandate of the national regulatory authority for pesticides. Within an organism's native range, many countries allow their use. Introduction and use of non-native organisms is generally regulated by a country's national plant protection organization following the *Guidelines for the export, shipment, import, and release of biological control agents and other beneficial organisms* (International Standard for Phytosanitary Measures 3, 2005) (IPPC, 2005). Requirements and procedures for risk assessment, import permits, documentation, etc. vary by country.

In countries such as Brazil, India and the USA, modified processes exist for the registration of biopesticides or for the registration of low-risk AS, whereas for many other countries the process for the registration of biopesticides is identical to that of synthetic pesticides. The registration requirements for botanicals are sometimes a grey area.

Some examples of the different approaches taken for the classification and regulation of biopesticides are as follows:

- **Brazil** has specific standards for the registration of biochemical biopesticides (Instrução Normativa Conjunta n° 32 – Produtos de Produtos Bioquímicos) (MAPA, 2005), biological control agents (Instrução Normativa Conjunta n° 2 – Registro de Produtos Biológicos) (MAPA, 2006b), microbials (Instrução Normativa Conjunta n° 3 – Registro de Produtos Microbiológicos) (MAPA, 2006c), and semiochemicals (Instrução Normativa Conjunta n° 1 – Registro de Produtos Semioquímicos) (MAPA, 2006a). These standards define what is covered by each group of biopesticides. Semiochemicals which are used for traps or for mating disruption (provided they are not applied to agricultural crops) are considered to require fewer restrictions. Information on potential benefits of the use is required to be included in the registration dossier for macrobials.
- In **Colombia**, a 'toxicological concept' from the Ministry of Social Protection is required for products that are microbial agents, entomopathogenic nematodes, entomoparasites, or plant extracts (MADR, 2002, 2011).
- The *Manual Técnico Andino para el Registro y Control de Plaguicidas Químicos de Uso Agrícola* (CAN, 2002), which is being applied in Ecuador, defines 'biological control agents' as 'natural or genetically modified agents that are distinguished from conventional chemical pesticides by their unique modes of action, for the smallness of the volume in which they are used and the specificity for the species in question

to fight'. Two types of biological control agents are listed: biochemical agents and microbial agents.

- **European Union** (EU) Member States are subject to the applicable EU legislation. The regulations for the placement of plant protection products on the market in EU countries are set forth in *Regulation (EC) No. 1107/2009* (EC, 2009). The EU does not distinguish biopesticides as a group separate from other conventional CPAs, but they are often grouped as 'low risk' products. In the EU, there is no official definition of what biopesticides are, but biopesticide categories are listed in supplementary law (guidance documents) for registration. Categories that are identified are semiochemicals, microbials and botanicals. Macrobials and semiochemicals (when used for monitoring purposes) are excluded from plant protection regulation. There are provisions for data requirements in EC Regulation 283/2013 (EC, 2013a) and EC Regulation 284/2013 (EC, 2013b). Some European countries fast-track biopesticides registration. In the EU, a class of products termed 'Plant strengtheners' is recognized as a distinct group. These are substances intended to protect plants against harmful organisms by activating the host plants' own defence mechanisms (induced resistance) or via competition with harmful organisms in the phyllosphere or rhizosphere for resources.
- **Italy** has reduced fees for biopesticides registration.
- The **Indian** Insecticide Act of 1968 does not provide a definition of 'biopesticides' or 'biological products'. There is a definition in the pending Pesticide Bill 2008. The website of the Indian pesticide regulator CIBRC (Central Insecticide Board and Registration Committee) (www.cibrc.in) defines three categories of biopesticides: (i) Microbial: NPV, EPF, Antagonistic bacteria/fungi and entomotoxic bacteria; (ii) Botanical: Neem, Pyretherum, herbal plant growth regulator, rotenone, eucalyptus etc.; and (iii) Pheromones. In India, biopesticides can be marketed under provisional registration 9–3B for 2 years before they obtain permanent registration 9–3. For Provisional registration information on toxicity to non-target organisms (other than humans) is not required but for permanent registration it is required. Macrobials are exempted from registration and so are pheromones until or unless they are to be used for mating disruption.
- In the **Philippines**, what is generally considered to be a biopesticide is included in definition of 'biorational pesticide'. In the Philippines, in principle biopesticides are registered under the same system as conventional CPAs, but within that system CPAs are tiered such that certain assessments are excluded for certain levels. The new regulatory body the Bureau of Agriculture and Fisheries Standards (BAFS), links biocontrol agents to organic certification, but also registration by the Fertilizer and Pesticide Authority (FPA) is needed. Products that are already registered but changing formulation do not need to renew organism data.
- The **South African** guidelines provided by the pesticide registrar defines 'biological remedy', 'bioproducts', 'biological products' and 'biopesticides' as 'any biological remedy, or any mixture or combination

of any substance or remedy intended or offered to be used for the destruction, control, repelling, attraction or prevention of any undesired microbe, alga, nematode, fungus, insect, plant, vertebrate, invertebrate, or any product thereof, but excluding any biological remedy in so far as it is controlled under the Medicines and Related Substances Control Act, 1965 (Act 101 of 1965), or the Hazardous Substances Act, 1973 (Act 15 of 1973) (Agricultural remedy as defined in Act 36 of 1947)' (DAFF RSA, 2015). South African guidelines from the pesticide registrar also define plant growth regulators, legume inoculants, biostimulants, biofertilizers, plant growth promoters and inoculants. Products covered by these definitions may contain microbials, macrobials, biochemicals such as plant growth regulators, semiochemicals, enzymes and hormones, plant extracts, legume inoculants and other inoculants such as free-living nitrogen-fixers, phosphate-solubilizing bacteria and plant growth-promoting bacteria. For microbials and macrobials, mass release permits from directorate Plant Health/Department of Environmental Affairs are a requirement for registration. Data requirements for registration are modelled on OECD guidelines. If the organisms are imported from another country, import permits and permission for commercialization are also required. For provisional registration of new AI, toxicological risk assessment reports from the registration authorities in Australia, the EU, Japan and the USA, together with a toxicological risk assessment, by an independent and accredited toxicologist, are accepted.

- **Spain** has reduced fees for biopesticides registration. The definition for *otros medios de defensa fitosanitaria* (OMDF) (other means of phytosanitary defence) set forth in Spanish regulations (Royal Decree 951/2014) (BOE, 2014) covers 'biological control organisms, traps and other devices which are not directly linked to pest control'. Substances that fall within the scope of Regulation (EU) No. 1107/2009 of the European Parliament (EC, 2009) and those whose inclusion in the list has been rejected by the European Commission are expressly excluded from the scope of application. Royal Decree 951/2014 outlines the procedures for commercializing OMDF. While OMDF do not require registration, commercialization of these products still requires advance notice to national authorities.
- **Thailand**'s data requirements for biopesticide registration have been aligned towards OECD and the EU since 2009.
- In the **USA**, the national pesticides regulator recognizes 'Plant incorporated protectants' as a group of biopesticides. Plant-incorporated protectants are defined as pesticide substances produced by plants, and the genetic material necessary for the plant to produce the substance. Thus genetically modified plants, for example via the introduction of the *Bacillus thuringiensis* protein toxin gene, would be included in this definition. In the USA, registration is not required either for macrobials or for pheromones that are used exclusively as attractants in traps.
- Malawi, Mozambique, South Africa and Tanzania are all member states of the **Southern African Development Community** (SADC), a regional

economic organization. SADC countries are working towards harmonizing their registration requirements for CPAs, including biopesticides.

The regulatory frameworks for biopesticides in several countries where tobacco is grown are summarized in Table 2.2. Other useful sources include: *ASEAN Guidelines on the Regulation, Use, and Trade of Biological Control Agents (BCA)* (Bateman *et al.,* 2014), FAO (2006), FAO and WHO (2014), Pesticides Board Malaysia (2016) and SENASA (1999).

2.2.2 Determining which biopesticides are allowed for use in a country

A market survey of biopesticides carried out in 2016 for 20 countries where tobacco is grown identified approximately 50 different biopesticide AS that are permitted to be used in tobacco (Bateman *et al.,* 2016). In a follow-up survey, leaf suppliers operating in even more countries indicated that they have experience with over 60 biopesticide AS, and would recommend 25 of those AS to others for use (Grunder and Bateman, 2017). Leaf suppliers indicate that there are many biopesticides that can help to replace HHPs (Bateman et al., 2017). Thus, it is clear that there is a wide range of biopesticides that can be used in tobacco, and many of these have already been demonstrated to be effective tools for use in IPM strategies, with the potential to support the phase-out of HHPs. That said, not all of these biopesticide AS are allowed for use in every country. Table 2.3 lists some of the biopesticides most widely recommended for use in tobacco, examples of the pests that they target and some of the countries in which they are registered for use in tobacco (as of 2017).

To find out the current registration status for tobacco biopesticides in a particular country, check that country's list of registered pesticides or registered biopesticides. As stated above in Section 2.2.1, some countries may exempt certain classes of biopesticides from registration requirements and so it is important to have a good understanding of the overall regulatory context for each country.

Background information and training material is provided for these biopesticide AS in Chapter 4.

2.3 Biopesticide Trial Guidelines

For any biopesticide efficacy trial that will be carried out, the project manager must develop a protocol describing all aspects of the trial, including introduction, objectives and a plan for data analysis. The information in this section can be used as a blueprint for the development of this protocol. The application of a uniform approach for trial protocol design will help to ensure that findings are meaningful and cross-comparable.

These biopesticide trial guidelines provide guidance for field trials designed to test the efficacy of CPAs against tobacco pests and diseases

Table 2.2 Overview of national regulatory frameworks for biopesticide registration in Argentina (AR), Brazil (BR), Colombia (CO), Ecuador (EC), Greece (GR), India (IN), Italy (IT), Malawi (MW), Mexico (MX), Mozambique (MZ), Pakistan (PK), Philippines (PH), Poland (PL), South Africa (ZA), Spain (ES), Tanzania (TZ), Thailand (TH), and the USA (US). (This information is current as of January 2018. Refer to national regulations for the most up-to-date information. Cells with '-' indicate that the information was not found in the regulations that were reviewed.)

Features of biopesticides regulatory frameworks	National regulations[1]: The columns below summarize how biopesticides are regulated in each country																	
	AR	BR	CO	EC	GR	IN	IT	MW	MX	MZ	PK	PH	PL	ZA	ES	TZ	TH	US
Are biopesticides defined in national regulations? (Y/N)	N	Y	Y	Y	N	Y	N	N	N	N	N	N	N	Y	N	N	Y	Y
Does the process for registration of biopesticides differ from the process of registration for conventional pesticides? (Y/N)	N	Y	Y	N	N	Y	N	N	N	N	N	Y	N	Y	N	N	Y	Y
Is information on efficacy required for registration? (Y/N)	Y	Y	Y	Y	Y	Y	Y	Y	Y	Y	Y	Y	Y	Y	Y	Y	Y	N
Is information on toxicity to humans required for registration? (Y/N)	Y	Y	Y	Y	Y	Y	Y	Y	Y	Y	Y	Y	Y	Y	Y	Y	Y	Y
Is information on toxicity to other non-target organisms required for registration? (Y/N)	Y	Y	-	Y	Y	Y	Y	Y	Y	Y	Y	Y	Y	Y	Y	N	Y	Y

Continued

Table 2.2 Continued

Features of biopesticides regulatory frameworks	National regulations[i]: The columns below summarize how biopesticides are regulated in each country																	
	AR	BR	CO	EC	GR	IN	IT	MW	MX	MZ	PK	PH	PL	ZA	ES	TZ	TH	US
Is information on environmental fate required for registration? (Y/N)	Y	Y	-	Y	Y	Y	Y	Y	Y	Y	Y	Y	Y	Y	Y	Y	Y	Y
Is information on the benefits of use required for registration? (Y/N)	-	Y	-	-	N	N	N	N	-	N	N	N	N	N	N	N	-	N
Is information for the assessment of quality required for registration? (Y/N)	Y	Y	Y	Y	Y	Y	Y	N	Y	N	Y	Y	Y	Y	Y	Y	Y	Y
Is information on the toxicity of additives required for registration? (Y/N)	-	N	N	Y	Y	N	Y	N	-	N	N	Y	Y	Y	Y	N	-	N
Is generic data based on strain or species which are already registered accepted for inclusion in registration dossiers? (Y/N)	-	-	-	-	-	Y	-	N	-	N	N	Y	-	-	-	N	-	-
Are biopesticide products containing certain AI exempted from the requirement of an MRL? (Y/N)	-	-	-	-	-	Y	-	N	-	N	N	-	-	N	-	N	-	Y

Continued

Table 2.2 Continued

Features of biopesticides regulatory frameworks	National regulations[1]: The columns below summarize how biopesticides are regulated in each country																	
	AR	BR	CO	EC	GR	IN	IT	MW	MX	MZ	PK	PH	PL	ZA	ES	TZ	TH	US
Are biopesticide products containing certain AI exempted from the registration requirement? (Y/N)	Y	N	N	Y	Y	Y	Y	N	-	N	N	N	Y	N	Y	N	Y	Y
Is there 'fast track' registration for biopesticides? (Y/N)	-	Y	-	-	N	N	N	N	-	N	N	-	N	N	N	N	-	-
Are fees reduced for the registration of biopesticides? (Y/N)	-	-	-	-	-	N	Y	N	-	N	N	-	-	-	Y	N	-	-
Are there other incentives in place for the registration of biopesticides? (Y/N)	-	-	-	-	N	Y	N	N	-	N	N	N	N	-	N	N	-	Y

Table 2.3 Widely recommended biopesticide active substances, the pests they target and examples of countries in which they are registered. (The information in this table is up-to-date as of September 2017. The list of countries where the biopesticide is registered for use is not exhaustive. There are potentially other countries where the biopesticide is registered.)

Biopesticide	Examples of targeted pests	Examples of countries where the biopesticide is registered for use
Ampelomyces quisqualis	Powdery mildew (*Erysiphe* spp.)	India
Bacillus firmus	Nematodes, e.g. *Meloidogyne* spp.	Greece, Italy, Mexico, Switzerland, USA
Bacillus pumilus	*Alternaria* spp., anthracnose (*Colletotrichum* spp.), *Rhizoctonia solani*, *Fusarium* spp., black shank (*Phytophthora* spp.), black leg (*Erwinia* spp.), blue mould (*Peronospora* spp.), *Sclerotinia* spp., frogeye (*Cercospora nicotianae*), powdery mildew, damping-off (*Pythium* spp.), *Sclerotium* spp.	USA
Bacillus subtilis	Many different pathogens, e.g. *Rhizoctonia solani*, *Fusarium* spp., *Pythium* spp., *Alternaria* spp., *Xanthomonas* spp., *Erwinia* spp., *Colletotrichum* spp.	Brazil, China, Greece, Turkey, USA
Bacillus thuringiensis	Lepidoptera, e.g. *Heliothis* spp., *Helicoverpa* spp., *Spodoptera* spp., *Manduca* spp., *Trichoplusia* spp., *Estigmene* spp.	One of the most widely registered biopesticides for tobacco. Countries include Argentina, Brazil, Ecuador, Greece, India, Indonesia, Italy, Mexico, Mozambique, Pakistan, Spain, USA, etc.
Beauveria bassiana	Many insect species such as aphids, whiteflies, thrips	Greece, India, Indonesia, Mexico, Mozambique, Pakistan, Thailand, USA
Coniothyrium minitans	*Sclerotinia sclerotiorum*	Greece
Garlic extract	Nematodes	Greece, Indonesia, Italy, Pakistan
Helicoverpa armigera nucleopolyhedrosis virus and (nucleopolyhedrosis) NPV for other Lepidoptera	*Helicoverpa* spp. (for HNPV) and other Lepidoptera species	Greece, India, Mexico, Spain, USA
Metarhizium anisopliae	Insects such as crickets and grasshoppers	Greece, India, Mexico, Mozambique

Continued

Table 2.3 Continued

Biopesticide	Examples of targeted pests	Examples of countries where the biopesticide is registered for use
Neem / Azadirachtin	A wide range of insects and even diseases	Widely registered globally
Paecilomyces lilacinus	Nematodes	Greece, India, Mexico, USA
Trichoderma spp.	A wide range of pathogens such as *Altenaria* spp., *Fusarium* spp., *Pythium* spp., *Rhizoctonia* spp.	Brazil, Greece, Italy, Mexico, Mozambique, Turkey, USA
Trichogramma spp.	Eggs of Lepidoptera	Most countries do not require registration of macroorganisms
Entomopathogenic nematodes such as *Steinernema carpocapsae*	Thrips, fungus gnats (*Bradysia* spp.), mole crickets (*Scapteriscus* spp.) and other insects	Most countries do not require registration of macroorganisms

under local conditions. They are intended to provide information on how to determine whether a particular chemical CPA or biopesticide is effective enough to be recommended to local farmers.

Topics covered in this section include:

- Objectives and scope of the trial guidelines
- Guidelines for Trial Protocol Design
- Tips for analysing data from biopesticide trials
- Economic analysis
- Trial protocol template
- Key references

2.3.1 Objectives and scope of the trial guidelines

The main objectives of the trial guidelines are:

- to standardize tobacco biopesticide trials across operations,
- to provide guidance on how to properly plan and manage CPA trials, how to collect data allowing for sound analysis, and using statistical tools so that valid conclusions can be drawn from the data.

The scope is as follows:

- This guideline provides guidance on how to carry out field trials. Wherever possible, it is advisable to build upon prior work done in the laboratory or in 'microcosms' (e.g. field cages).

- Throughout the guideline the generic term 'CPA' is used, but the main focus of this guideline is to provide guidance on how to design biopesticide trials. In most cases, trials of chemical CPAs and biopesticides require similar set-ups in terms of experimental design, but special attention may be needed when testing living biological control agents, as they may have special requirements, such as using the correct plot size for biocontrol agents that are able to disperse long distances. Where relevant, specific information is provided to account for any special requirements that may need to be applied to the testing of biopesticides.

- This guideline provides details on how to conduct trials at one given place, such as a research centre. It does not directly refer to trials conducted in farmers' fields although many of the details provided here would also be valid for research conducted with farmers.

- This guideline may be used for trials to be conducted in the seedbed or the field. The only major differences between these settings are plot size and the distance between plots.

- It is assumed that for most situations the test products are already registered in the country, though this may potentially be only for other crops. For products not yet registered in the country, but which appear to be promising, national regulations must be consulted to ensure compliance with applicable legislation.

- This guideline should by no means be considered to be fully comprehensive, considering all details and all possible scenarios. For additional aspects/details on conducting CPA trials, refer to the documents listed in the references, especially the references from the European Plant Protection Organization (EPPO) which has substantial expertise in that area and has published relevant information on the subject, even though generally on a more generic level (EPPO, 1997, 1999, 2012a, 2012b, 2012c, 2004, 2014; see also Candolfi *et al.,* 2000). For more information on specific technical terms (e.g. blocking, choosing treatments), refer to the training material which is available at the Tobacco IPM toolbox website (https://www.tobaccoipm.com).

2.3.2   Guidance for trial protocol design

This section provides detailed guidance on how to design a CPA trial protocol. The main features will be: (i) front page; (ii) background; (iii) objectives; (iv) trial design; and (v) technical applications. A template for filling in the CPA trial protocol is provided in Section 2.3.4.

- **Front page**

   This should include the project name, location, project leader and team, crop year, crop type/variety (this should be the one that is most widely used) as well as revision history (version number, author, effective date (mm/dd/yyyy)). A unique trial code per country can also be added.

- **Background**

  This section should introduce the problem and provide relevant background information. It is recommended that a literature research is carried out to assess the current state of knowledge and to make sure that sufficient information is collected for the products in question.

- **Objectives**

  This section should set out clearly why the study is to be performed. For instance, the trial may be being planned to see whether the product can be successfully embedded in a crop protection scheme such as an IPM programme, to test its compatibility with biological agents or to test for any effect it may have on crop quality, and so on. At this stage, information should also be provided on the project boundaries, e.g. whether it is for a specific region only, a specific group of farmers, or a particular variety.

  Difficulty with study design often comes down to poor objectives. To avoid this, the objective(s) should be stated clearly. Is the aim simply to compare different CPAs with each other? Is it about testing efficacy compared with one particular reference product? Is it about using different dosages of one or several CPAs? Is it about cost efficiency? It may also be worth stating what level of effect size (treatment differences) would be considered meaningful as this will have implications for the experimental design, for example the number of replicates that will be needed. Having clear objectives with set criteria such as these will facilitate decision making after the trial has been completed. Thus, it is important to be as clear, complete and precise as possible when stating the objective(s). The overall objective (aim) of the study followed by specific objectives should be listed.

  If possible a trial series should be carried out, in which trials of the same design and treatments would be conducted at several locations or in two consecutive years. This would generate more data for statistical analysis, greatly enhancing the power and validity of the conclusions.

  It may be of interest to compare a 'full IPM system' with any 'conventional practice' currently in use. This type of trial would obviously include both seedbeds followed by the field. Such trials are resource intensive but may yield important insights. This guideline can also be applied in these situations. Care would then be needed to transplant correctly from seedbed IPM plots into field IPM plots.

- **Trial design**

  Choosing treatments

  The treatments are to some extent predefined by the objectives. However, care should be taken to include a reasonable number of treatments; ideally, this should be between two and ten treatments. Low numbers of treatments will provide limited data and there is a risk that the results will not be statistically significant. Conversely, using too many treatments can create excessive data, making statistical analysis and interpretation difficult. It is also important to ensure that the number of replicates per treatment is sufficient to maintain the statistical power and quality of conclusions.

Normal biological variation, researcher bias and environmental variation are all factors that can skew data, and controls are therefore always needed. These controls provide a baseline against which the treatments are compared. Different types of control exist and the choice of which to use will depend on the specific objectives, however, the following should be considered:

- A **negative control** basically means doing nothing to the plot in question. In most cases this should be a so-called procedural control in which no CPA is applied, but the same amount of water is used, possibly together with the same solvents used in the other treatments. In addition to providing a baseline, one purpose of such untreated control is to demonstrate the presence of an adequate pest infestation. If there is only a low level of pest population observed in the untreated control, efficacy cannot be demonstrated and results will not be meaningful.
- A **positive control** may be a product whose efficacy is already known. This could be a CPA that has been successfully used for many years to control the pest in the past. If the biopesticide test products show no significant reduction in the pest populations, the trial should at least be able to demonstrate that the method has worked by showing that the positive control still had an effect on pest control.

Second factor

The trial design can also include a second factor. In this case, the treatments are defined by combinations of two or more treatment factors each at two or more levels. This will inevitably increase the number of plots in the study, and thus resources required. However, including a second factor will increase the amount of information obtained and enable specific interactions between the two factors to be identified – such as whether the difference in pest reduction between CPAs depends on the planting density (the second factor here). Which additional factor to include and whether to use it, will depend on the objectives. For instance, products of interest can be tested at both a high dose and a low dose, with two different nozzle types, or with varying amounts of water.

Testing three products plus a control, together with two different varieties of tobacco (the second factor), would require $4 \times 2 = 8$ treatments. If the factors interact, then this trial design would be essential. However, if there is no interaction between the factors being tested then such a design would still be efficient as it will provide the same information and with the same precision as a similar trial investigating the factors separately.

- **Technical application**

  The application method and equipment used must be explained thoroughly, including:

- The doses used (kilograms or litres formulated product per hectare), and the spray volumes (and, if relevant, any significant deviations from the intended dosage). There may be incomplete knowledge and, particularly for biopesticides, special care is needed to collect and consider all relevant information. This may potentially require involving international experts.
- The operating conditions, in so far as they may affect efficacy or selectivity (e.g. for sprays, pressure, nozzle type, spray quality and speed of travel of sprayer).
- The number of applications to be made and dates of application.
- The growth stage of the crop (and for herbicides, the weeds) at the time of each application; see BBCH Growth Stage Keys (Meier, , 2001) and CORESTA Guide No. 7 (CORESTA, 2009).
- Where appropriate, the development stage of the pest or infestation level at the time of each application.
- Any notes about the system that are important for the timing of applications (calendar, phenological stage of crop, threshold levels or development stage of pest, external warning system).

  If possible, the equipment used for the trial should be the same as the equipment used by the majority of farmers. In some instances, it may be advisable to use smaller-scale application methodology for field trials if this makes it easier to conduct the appropriate number of replicates. All test equipment should be calibrated and documented before each trial to ensure the accuracy of application rates.

  It is important to apply the treatments at the same time as much as possible, a problem that can arise in very large trials with many treatments. Rational pesticide use should be strongly considered, such as wearing appropriate protection equipment, or avoiding application during windy conditions.

  The same equipment should be used for all different treatments as much as possible. In the case of several CPA applications within the trial, the same staff should carry out the spraying of all plots each time. If this is not feasible then the applicators must be randomly assigned to the treatments.

Note Specialized equipment may be needed for biopesticides, e.g. a higher volume of water may be required to give the living organisms a better chance of survival and to increase their efficacy.

Note In the case of application of living organisms, take extra care to note product shelf life and environmental conditions (e.g. temperature, moisture).

Note If the experiment involves the application of a living biocontrol organism or a treatment based on semiochemicals (e.g. mating disruption), this will involve a much bigger plot area and thus will usually require totally different equipment.

2.3.2.1 General design/randomization

Randomization is a key requirement in statistical data analysis. In order to ensure randomization, the most commonly used designs for CPA field trials are the Completely Randomized Design (CRD) or the Randomized Complete Block Design (RCBD).

2.3.2.1.1   CRD.   This type of trial can be used where conditions are relatively homogeneous. It is a simple design and is specifically recommended when using a relatively small number of replicates, because it does not reduce the degrees of freedom in statistical data analysis, as does the RCBD. Plots are placed at random throughout the field being used for the trial. Randomization should be carried out by using an Excel number generator, or similar online random number generator. As stated above, the assumption is that the trial is conducted at one site, usually a research centre or similar. Thus, in many cases, the trial will be implemented on a scale of less than 1 ha or not much more. Under these assumptions, it should be possible in most cases to choose a field that is rather homogeneous regarding soil conditions, previous crop, slope etc. See Fig. 2.1 for a suggested CRD design.

2.3.2.1.2   RCBD.   If the field to be used is very large and it is suspected or known that there is some environmental variation across the site (site heterogeneity), then a block design can be used to remove the effects of this variability. For instance, soil nutrient status may be better in the northern part of the field, gradually getting worse towards the south. In this case blocks would be created from north to south. The layout of the blocks therefore requires some preliminary knowledge of the trial area.

RCBD is the simplest form of a randomized block design, in which all treatments occur once in each block; and treatments are placed randomly within the block. Since each block contains a complete set of treatments, any differences found between blocks are not due to the treatments, and this variability can be estimated as a separate source of variation. Note that plots within one block do not need to be next to each other, plots must just be combined in blocks showing similarities with respect to the environmental variations being controlled for. See Figs. 2.2 and 2.3 for suggested RCBD layout designs.

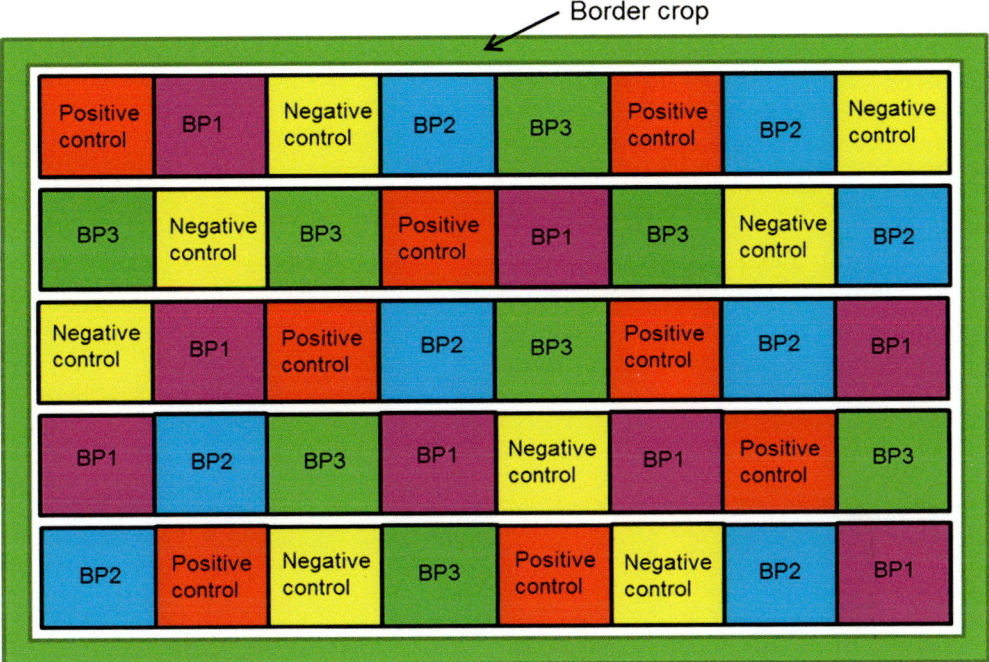

Fig. 2.1 Example of a simple Completely Randomized Design with five treatments (BP1-3, negative and positive controls), and eight replicates. The positive control may be a conventional CPA; the negative control indicates no CPA application and three different biopesticides are tested. Due to the relatively high number of replicates, a second year may not be necessary.

2.3.2.2   Site selection

- A homogeneous site is ideal; consider what the land has been used for during the preceding season, e.g. avoid a piece of land where two or more crops with very different inputs have previously been grown.
- The site chosen must show a certain level of pressure from the pest under consideration or no calculation of efficacy will be possible at all. Artificial infestation with the pest in question should be avoided if possible, but in some cases it may be required to guarantee a certain level of pest incidence to make the trial meaningful.
- Note down the key information from the site (soil type, slope, previous crop, etc.) that could potentially help to interpret data collected later on.
- For the biopesticide trials considered here, fields with a known history of use of persistent pesticides should be avoided.

2.3.2.3   General agronomic practices

It is very important that the complete trial area is treated the same way, except for the CPA applications. This not only includes measures such as plant

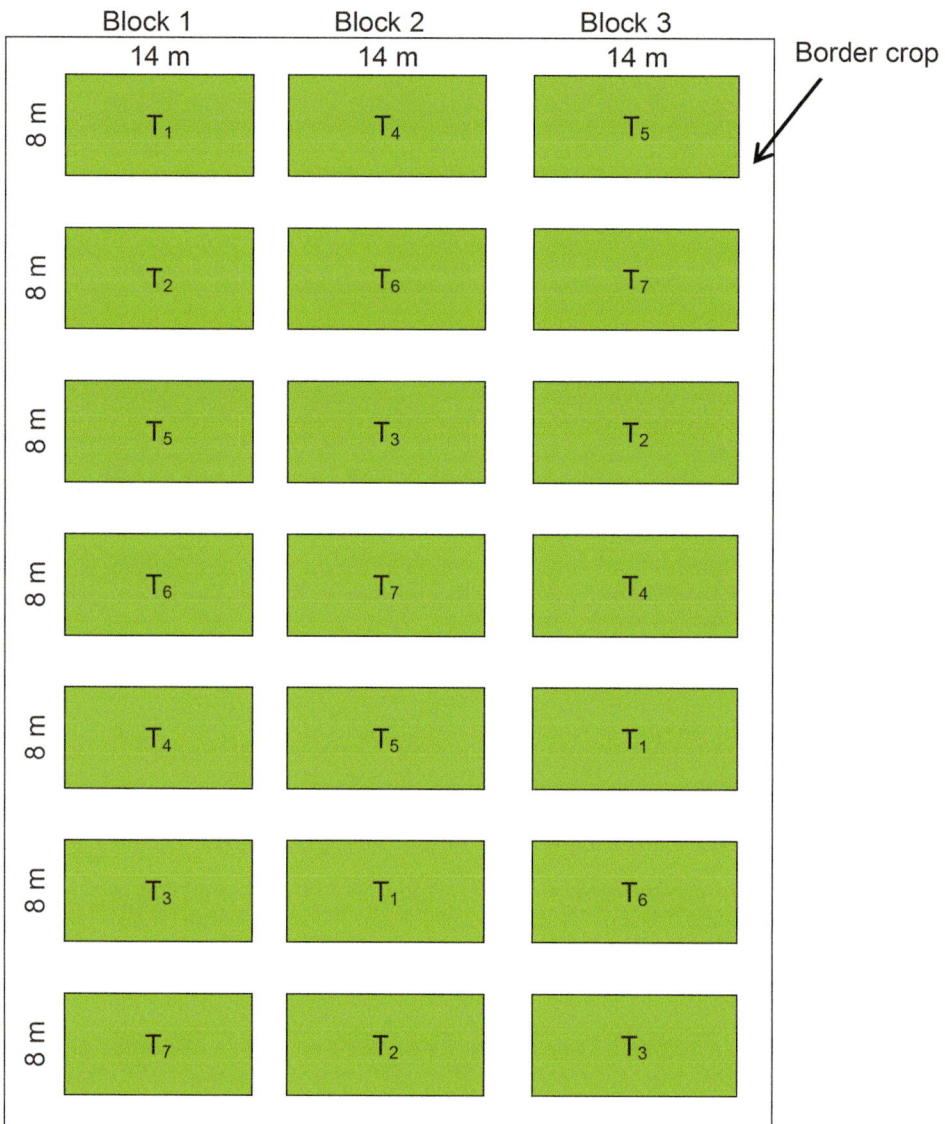

T=treatment

Fig. 2.2 Example of a Randomized Complete Block Design with three blocks, and seven treatments for one replicate randomized within each block (thus n=3). This should be repeated for a second year to generate enough replicates for statistical analysis.

spacing, fertilization and weeding, but also the management of pests other than those targeted by the products to be tested. An example may be in the case of CPAs being tested against blue mould, but at the same time problems with aphids are being observed in some of the plots. Unless this is causing problems with the specific objectives of the trial, all plots should

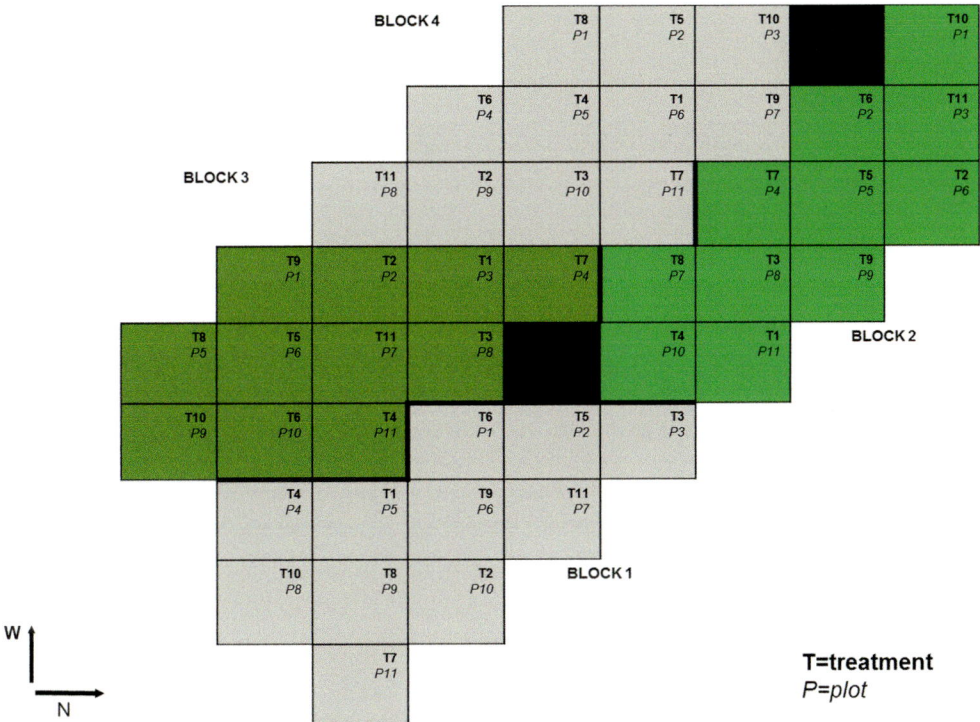

Fig. 2.3 Example of a RCBD with blocks of irregular shape (11 treatments, n=4), with plots of 10×10 m. The two black plots are not being used as part of this trial due to the presence of too many stones. These two plots should also be excluded from future trials on this land.

be treated against aphids, including plots not affected by aphids. Treating all plots uniformly will mean that data collected will increase substantially whilst not adversely affecting the precision of the trial. All measures should be typical for the region and agreed with the local agronomy team. All agronomic measures must be clearly reported.

2.3.2.4 Plots

Plot size should be no less than 50 m² but ideally more than 100 m². However, if there are constraints on the space available, then using a greater number of smaller plots will be more powerful in statistical terms than fewer larger plots. The distance between plots should be no less than 2 m, to avoid edge effects (i.e. spraying a CPA on one plot may affect the neighbouring plots). The simplest way to achieve this would be to account for the row distance used in that particular variety (e.g. in tobacco with a 1.2 m distance between rows, the distance between plots should be 2.4 m). The rows located between the treated plots can and indeed ought to be planted with tobacco. If trials are conducted in the seedbed, plot size and distance between plots will have to be reduced, but care is still needed to avoid edge effects. This may be addressed by, for example, using smaller scale spraying equipment.

If living organisms are tested, these distances will need to be adapted to account for dispersal of the agents. Also, if the trial is likely to last longer than one pest generation or if pest dispersal is high during the trial period, care is needed and distances between plots should be increased. At least five rows of tobacco should be grown around the trial area. Make sure that the plots' size and shape conforms to the size of the agricultural equipment used (e.g. boom size). In general, elongated plots are recommended, e.g. 6×20 m. All plots should be very clearly labelled.

Note A larger plot size may be needed for application of biopesticides, particularly for living biocontrol organisms or a treatment based on semiochemicals (e.g. mating disruption). This is because in both cases the dispersal capacity of the insects involved is substantial. Therefore, plots of not less than 1 ha should be selected, with distances of at least 100 m between plots. These design requirements would mean that it is unlikely the trial will be able to be conducted at one single research centre. If fields of these sizes are not available, then the objectives should be reconsidered or at least use fields that are 'as large as possible'.

2.3.2.5   Subplots (within plot measurements)

Measurements may be taken at several locations in the main plot, known as 'subplots'. As an example, specific measurements may be taken from ten plants each in five subplots (located in the four corners and the centre) of the main plot. One variable that will often change a lot within plots is pest incidence, so it is advisable to have several randomly selected subplots from which information on pest incidence is collected. The number of subplots is suggested to be five.

Note The data collected from subplots are not independent data and can therefore not be treated as 'replicates'. Instead, values obtained from subplots should be averaged, resulting in a measurement of higher quality for the main plot.

2.3.2.5.1   REPLICATES.   Replicates are key for reaching a certain level of statistical power in the trial and to allow strong conclusions to be drawn. This statistical power depends on three parameters: variability of data, effect size (the treatment differences of interest), and replication. In general, having more replicates gives more precision in the data and thus increases the likelihood of finding differences between treatments. If treatment effects are missed as a result of using too few replicates, all the resources put into the trials will have been wasted.

In addition, a higher number of replicates may act as kind of insurance, i.e. may compensate for some plots that have to be removed from analysis, and thus may avoid losing the whole trial. It is suggested that, for a one-factorial trial (where only one factor is being studied), the number of replicates is at least four. A higher number of replicates will be necessary when:

- relatively small differences between CPAs, or small differences are anticipated;
- a high variability is expected; and/or
- a small number of treatments are involved and the trial is blocked.

It is strongly recommended that trials are conducted over a minimum of two growing seasons or years to capture any variation between seasons that could affect the efficacy of the product. 'Year' would then be another factor for the statistical analysis, but assuming the same protocol is being followed, all data collected from the second year can be considered true replicates and should be accounted for when planning the number of replicates. Therefore, if three replicates are conducted in year 1, and another three replicates in the second year, the total number of replicates would be six, which then is in agreement with this guideline.

2.3.2.6 Data collection

2.3.2.6.1 MEASUREMENTS. The type of measurements to be taken depends on the objective, but an obvious one to include is pest incidence. Yield and leaf quality are other important measurements. For yield assessment, make sure that the plot size is measured accurately, as even small differences in plot size among treatments will bias the results. If chemical CPAs are involved in the trial, residues should be also assessed.

Following CPA applications, a check for phytotoxic effects is obligatory. Furthermore, an assessment of key non-target effects, particularly natural enemies playing a role in pest control, is highly advisable. Additional measurements that may be considered include any negative effects on succeeding crops that may indicate an effect on the sustainability of the production system.

All crop husbandry measures should be written down including sowing, transplanting, fertilization, etc. It is important to take measurements on abiotic conditions, such as temperature, precipitation, extreme weather events, etc. These may help later on when interpreting the results. All costs relevant to the application of the test product should be noted.

Further points are listed here to assist in the correct assessment of pest incidence, a key parameter in CPA efficacy trials.

- Target pests (including pathogens and weeds) should be identified to species level wherever possible. If there are difficulties in identifying certain organisms, the submission should outline the characteristics of the pest, disease or weed and/or the classification system used in attempting to identify it or distinguish it from other related organisms that are present. A counting or measurement system will need to be developed and described for pests that are difficult to see, such as thrips. Generally the best is to provide exact counts, e.g. 12 individuals of a given species may be found in a given plot. Information can also be provided on factors such as number of leaves/number of plants damaged/infested, in which case a scale is needed on the level of severity of that attack (see

'Visual estimations' below). Such a scale must be established before the start of the trial.

- The number of pest incidence assessments to be carried out is difficult to generalize because it depends on the life history of the pest in question, as well as on the mode of action of the CPA. However, data on the pest in question should be recorded both shortly before treatment (e.g. 24 h) and again shortly afterwards (24 h). Thereafter, data on pest incidence may be collected again after 72 h and 7 days. Subsequent collection of pest incidence data will then depend on the specific needs of the trial. For biopesticides, the general advice is to wait at least 1 day after application before collecting data. This is because time is usually needed for the biopesticide to kill the pests and for these effects to be observed.

The data collection scheme needs to be followed for each CPA application.

2.3.2.6.2 GENERAL POINTS. Where possible, take measurements of continuous variables (e.g. weight, size) and avoid taking discrete ones (e.g. counts, especially when numbers are small), or even categorical ones (e.g. presence/absence). This is because statistical data analysis is more powerful when using continuous variables.

To avoid the risk of inconsistent data collection, make sure that the personnel taking measurements are well trained on the subject and that they note their name on the data sheet. Ideally, the same staff should be involved in taking measurements throughout the trial and changes avoided as much as possible. Any changes in staff collecting data should be recorded. The use of an ink pen is recommended for data gathering. Any errors in the process of data gathering will increase the variability of the data set, resulting in reduced precision in the analysis and potentially limiting the conclusions that can be drawn.

'**Blinding**': Experimenters might be prone to bias, i.e. they may tend to judge differently depending on what treatment they look at. To avoid such bias, 'blinding' is recommended, meaning that the person taking the measurements (on yield, pest incidence, etc.) does not know on which plot a control or any other treatment has been applied. The plots will still need to be clearly labelled but this can be done by using a code to refer to the treatments. This procedure is advisable as it avoids any biased outcomes. The same measurement method should be used in all experimental units, e.g. observe all plots from the same direction to avoid differences in lighting.

Visual estimations: These determine values subjectively. Examples of visual variables are weed cover or numbers of lesions on a leaf. In general, an estimation is easier to make by using a comparison with a reference or an untreated control than it is in absolute values. To ensure the accuracy of such estimations, the observer should be trained to make the estimations and their observations should be calibrated against a standard.

2.3.2.7 Report

2.3.2.7.1 STATISTICAL ANALYSES. Data on yield and pest incidence should always be submitted to statistical analysis. Additional factors such as quality, phytotoxic effects or non-target effects may also be analysed statistically based on the objectives of the trial. If blocking was done, this needs to be recorded in the analysis as an additional factor. Similarly, the year (if the trial was conducted over more than 1 year) or site (if the trial was conducted at multiple sites) should also be incorporated into the analysis.

The statistical software recommended is Minitab 16 (http://www.minitab.com), but other software may be used if preferred. A copy of this program can be requested locally.

2.3.2.7.2 REPORTING. After the CPA trial has been completed, a report must be issued with a summary of the procedures (considering all relevant points of this guideline), the results (derived from statistical analyses), the conclusions and recommendations.

2.3.3 Tips for analyzing data from biopesticide trials

The aim of this section is to provide some specific tips, adding to and partially re-emphasizing points mentioned in the guideline above. It is first important to have a thorough look at the guideline text and, if any aspect is not fully clear, to consult with the corresponding training material available at the Tobacco IPM toolbox website, or other experts in the field.

- From the start, consider that planning for data analysis is also part of planning for the trial. It is therefore important to know what data are to be collected and how these data should be analysed. This should ensure that data are collected correctly. Ultimately, the data collected and analysed should be able to support the objectives (the question the trial is seeking to answer).
- No particular software is needed and simple designs can even be analysed with MS Excel.
- Check for outliers. If an outlier is suspected, check the following: Was the value typed in correctly? Was there an experimental problem with that value? Is it due to biological diversity? If the outlier is due to a mistake remove it from the analysis, if it is due to chance, keep it in the data set.
- Consider transforming the data before doing statistical analysis if distribution is not normal or variances are not similar among treatments. Transformations commonly used are based on log, square-root or arcsine. When reporting, always use the original data.
- To avoid over-complication, data from different collections over time should be analysed separately, e.g. data collected 24 h after treatment should be analysed separately from data taken 1 week after treatment.

- Check whether the results of the trial are coherent, and specifically whether the reference product (positive control) gave the expected result in comparison with the untreated control. If this not the case, then there might be unknown factors affecting the results and care should be taken when making any conclusions.
- Provide statements on the level of statistical difference (i.e. refer to the p-value), but also provide statements on the magnitude of any given effect, e.g. the proportional increase or decrease in pest populations.
- The statistical methods used for analysis of the data must be reported in sufficient detail; also make sure to report not only the p-value but also the test statistic (e.g. the F-value in an ANOVA) and the degrees of freedom.
- Taking subsamples from a plot is recommended for variables with higher variability, such as pest incidence. When analysing the results remember that, as stated earlier, data taken from several locations of the plot are not independent data and cannot be treated as 'replicates'. Subsample data should be averaged, resulting in a measurement of higher precision for the main plot.
- Variability is a major issue for statistical analysis in general and should be minimized as much as possible. Any aspect of the trial that can be controlled by the experimenter should be considered very carefully. For example, for yield assessment, make sure that plot size is measured accurately, as even small differences in plot size among treatments will increase the variability of the data and decrease the precision of the analysis.

2.3.4   Trial protocol template example

A template for filling in the CPA trial protocol is given below. This should be filled in for each CPA trial and submitted for approval.

1 Front Page
1.1 Project name
1.2 Name of project leader and team
1.3 Crop year and data
1.4 Selected site
1.5 Work/field unit identification
1.6 Approvals

2 Current Status and Objectives
2.1 Situation assessment / background
2.2 Objective
2.3 Scope

3 Methodology and Materials
3.1 Trial design, trial size, no. of replicates
3.2 Treatments and application method
3.3 Field management e.g. variety, fertilization, plant spacing
3.4 Plot identification
3.5 Data collection
3.6 Harvesting dates

4 Costs
5 Timelines
6 Accountabilities

2.4 Guidelines for Economic Analysis of Biopesticide Trials

The primary objective of conducting field trials with an alternative product is to determine whether the treatment has a significant effect on the crop, especially in terms of successfully controlling a pest, improving its yield or its quality. However, in order to be adopted by farmers, a biopesticide must also fulfil several criteria going beyond its efficacy (see decision matrix in Section 2.1.2). Very often, as for other technologies, socio-economic factors play an important role in this regard.

An economic analysis of the trial data is an important stage that should be conducted to determine if an alternative biopesticide product is a convincing economic choice. Several methods can be used to define the economic performance of a new technology. Where one specific crop is under consideration, partial budget analysis, which is commonly used to compare the economic benefits of technologies, can be applied. This methodology quantifies and compares the costs and revenues that are affected by a potential change of practice or through the introduction of a proposed technology and helps to determine the profitability of that change. This will then assist in the process of deciding whether or not to adopt the proposed technology.

Partial budget analysis does not examine all of the production costs, but only those that are affected by the change, i.e. the costs that will vary between the two technologies or practices tested. On a hectare basis, this will include purchased inputs, labour and machinery. In the case of a comparison between two CPAs for the same crop, the analysis can be simplified to a comparison of the differences between costs of production and revenue linked to each product. In the case of on-station trials, caution should be taken with regard to the evaluation of revenue and costs. Firstly the yield has to be adjusted to a certain degree to prevent an overestimation of the revenue that farmers are likely to obtain from a treatment. This is because experimental yields are usually higher than yields farmers can expect under normal conditions. Published literature suggests that the yield can often vary between 5% and 30% (sometimes more) depending on the conditions of the trials and the farmers' individual practices (CIMMYT, 1988; Hall *et al.,* 2013; Laidig *et al.,* 2017). The need for an adjustment to take this into account will be determined by the research team, based on the conditions of the trials. Farmer's fields with similar cropping conditions and in this case using the same conventional CPAs can be used as a basis for evaluating the yield difference compared with experimental fields, and this can be used as the factor to be applied to adjust the yield accordingly. Labour costs will also need to be estimated based on the time farmers and farm workers will have to spend in their field to conduct the required work.

Table 2.4 Example of partial budget for two CPAs.

	Conventional CPA (1)	Biopesticide (2)
Adjusted tobacco yield (Y)	Y_1	Y_2
Tobacco buying price (P)	P	P
Revenue (A = Y x P)	A_1	A_2
Costs of the CPA (+water)	$C_{1.1}$	$C_{2.1}$
Costs of labour to apply CPA	$C_{1.2}$	$C_{2.2}$
Costs of machinery to apply CPA	$C_{1.3}$	$C_{2.3}$
Any extra costs	$C_{1.i}$	$C_{2.i}$
Total costs that vary (C_{tot})	C_{tot1}	C_{tot2}
Net Benefit (B = A – C)	B_1	B_2

2.4.1 Decision rule

As highlighted in the decision matrix in Section 2.1.2, certain key points must be taken into consideration when selecting biopesticides. Regarding the economic aspects, the first condition to fulfil is that the net benefit of their use should be positive, i.e. the revenue should be greater than the associated costs:

- $B_2 > 0$

Comparison of the net benefits:

- $B_2 > B_1$
  While comparing the two alternatives, if the net benefit of the biopesticide (B_2) is greater than that from the conventional CPA (B_1), the biopesticide can be recommended to farmers.
- $B_2 < B_1$

  If the net benefit of the biopesticide (B_2) is smaller than that from the conventional CPA (B_1), further consideration is needed, especially regarding any non-monetary benefits. Other advantages of the technology, positive externalities (consequences of an action that positively affect unrelated third parties and whose impact is not reflected in the market price of the action, e.g. positive effect of the adoption of biopesticides on the environment) together with any farmer preferences should also be considered in the decision-making process. For example, the environmental benefit resulting from more environmentally friendly practices, any health aspects or the reduction of residues and thus reduction of business risks are valuable outcomes that can result from the use of biopesticides. The reduced PHI and the impact on beneficials are other examples of the advantages of biopestides compared with conventional CPAs. When deciding what would constitute an acceptable difference in the net benefit between the two technologies $(X = B_1 - B_2)$ all these aspects should be taken into consideration. The cost of adoption (e.g. learning about the new technology) which could prevent the farmer adopting a new technology should also be taken into account, particularly in cases where similar net benefits are found. Once the positive externalities and other advantages of biopesticides have been evaluated and the decision taken to promote that biopesticide, some thought should be given to the incentives that will be needed to promote its use.

References

Bateman, M., Grunder, J., Holmes, K., Babendreier, D. and Jenner, E. (2016) *Commercially available biopesticides for use against pests of tobacco: Findings from a review of 20 countries' lists of registered crop protection agents*. CAB International, Wallingford, UK.

Bateman, M., Bateman, S. and Grunder, J. (2017) *Report on identification and assessment of highly hazardous pesticides and their alternatives: Findings from a review of crop protection agents for pests of tobacco in 19 countries*. CAB International, Wallingford, UK.

Bateman, R., Ginting, S., Moltmann, J. and Jäkel, T. (2014) *ASEAN Guidelines on the Regulation, Use, and Trade of Biological Control Agents (BCA)*. Prepared by the Regional BCA Expert Working Groups on Application and Regulation. Association of Southeast Asian Nations, Jakarta.

BOE (2017) *Real Decreto 534/2017, de 26 de mayo, por el que se modifica el Real Decreto 951/2014, de 14 de noviembre, por el que se regula la comercialización de determinados medios de defensa fitosanitaria*. Boletin Oficial del Estado 126, 42959–42962. Madrid, Agencia Estatal.

CAN (2002) Resolución 630 In: *Manual Técnico Andino para el Registro y Control de Plaguicidas Químicos de Uso Agrícola*. Gaceta Oficial, Ecuador. Comunidad Andina de Naciones (CAN), Lima, Peru.

Candolfi, M.P., Blumel, S., Forster, R., Bakker, F.M. and Grimm, C. (2000) *Guidelines to evaluate side-effects of plant protection products to non-target arthropods. IOBC, BART and EPPO Joint Initiative*. IOBC/WPRS, Ghent.

CIMMYT (1988) *From Agronomic Data to Farmer Recommendations: An Economics Training Manual. Completely revised edition*. CIMMYT, Mexico DF, Mexico.

CORESTA (2009) *Guide No. 7. A scale for coding growth stages in tobacco crops*. Cooperation Centre for Scientific Research Relative to Tobacco, Paris.

DAFF RSA (2015) *Guidelines on the Data Required for Registration of Biological/Biopesticides Remedies in South Africa*. Department of Agriculture, Forestry and Fisheries, Johannesburg.

EC (2009) *Regulation (EC) No. 1107/209, concerning the placing of plant protection products on the market*. European Commission, Geneva.

EC (2013a) *No 283/2013 of 1 March 2013 setting out the data requirements for active substances*. European Commission, Brussels.

EC (2013b) *No 284/2013 of 1 March 2013 setting out the data requirements for plant protection products*. European Commission, Brussels.

EPPO (1997) Phytotoxicity assessment. EPPO Standards – Efficacy evaluation of plant protection products – No PP 1/135(2). *OEPP/EPPO Bulletin* 27(2), 389–400.

EPPO (1999) Design and analysis of efficacy evaluation trials. EPPO Standards – Efficacy evaluation of plant protection products – No. PP 1/152(2). *OEPP/EPPO Bulletin* 29(3), 297–317.

EPPO (2004) Efficacy evaluation of plant protection products – No. PP 1/181(3). *OEPP / EPPO Bulletin* 34(1).

EPPO (2012a) Design and analysis of efficacy evaluation trials. *EPPO Bulletin* 42(3), 367–381.

EPPO (2012b) Conduct and reporting of efficacy evaluation trials, including good experimental practice. *EPPO Bulletin* 42(3), 382–393.

EPPO (2012c) Number of efficacy trials. *EPPO Bulletin* 42(3), 405–408.

EPPO (2014) Principles of acceptable efficacy. No. PP 1/214(3). *EPPO Bulletin* 44(3), 274–277.

FAO (1995) The farming systems approach to development and appropriate technology generation. Available at: http://www.fao.org/docrep/v5330e/V5330e0h.htm (accessed 12 September 2017).

FAO (2006) *International Code of Conduct on the Distribution and Use of Pesticides*. Food and Agriculture Organization, Rome.

FAO (2016) *Guidelines on Highly Hazardous Pesticides. International Code of Conduct on Pesticide Management*. Food and Agriculture Organization, Rome, Italy.

FAO (2017) Registration Toolkit. Available at: http://www.fao.org/pesticide-registration-toolkit/tool/home/ (accessed September 2017).

FAO and WHO (2014) *International Code of Conduct on Pesticide Management*. Food and Agriculture Organization, Rome, Italy.

Grunder, J. and Bateman, M. (2017) *Report on leaf suppliers' experiences with biopesticides*. CABI, Wallingford, UK.

Hall, A.J., Feoli, C., Ingaramo, J. and Balzarini, M. (2013) Gaps between farmer and attainable yields across rainfed sunflower growing regions of Argentina. *Field Crops Research* 143, 119–129.

IPPC (2005) *Guidelines for the export, shipment, import, and release of biological control agents and other beneficial organisms (International Standard for Phytosanitary Measures 3, 2005)*. International Plant Protection Convention, FAO, Rome.

Laidig, F., Piepho, H.P., Rentel, D., Drobek, T., Meyer, U. et al. (2017) Breeding progress, environmental variation and correlation of winter wheat yield and quality traits in German official variety trials and on-farm during 1983-2014. *Theoretical and Applied Genetics* 130(1), 223–245.

MADR (2006) *Resolución 187 de 2006 por la cual se adopta el Reglamento para la producción primaria, procesamiento, empacado, etiquetado, almacenamiento, certificación, importación, comercialización, y se establece el Sistema de Control de Productos Agropecuarios Ecológicos*. Ministerio de Agricultura y Desarrollo Rural, Bogotá, Colombia.

MADR (2011) *Resolución 698 de febrero de 2011 por medio de la cual se establecen los requisitos para el registro de departamentos tecnicos de ensayos de eficacia, productores e importadores de bioinsumos de uso agricola y se dictan otras disposiciones*. Ministerio de Agricultura y Desarrollo Rural, Bogotá, Colombia.

MAPA (2005) *Instrução Normativa Conjunta n° 32, de 26 de outubro de 2005: Produtos de Produtos Bioquímicos*. Secretaria de Defesa Agropecuária, Agência Nacional De Vigilância Sanitária, Instituto Brasileiro do Meio Ambiente e dos Recursos Naturais Renováveis. Ministério da Agricultura, Pecuária e Abastecimento, Brasilia.

MAPA (2006a) *Instrução Normativa Conjunta n° 1, de 23 de janeiro de 2006: Registro de Produtos Semioquímicos*. Ministério da Agricultura, Pecuária e Abastecimento, Brasilia.

MAPA (2006b) *Instrução Normativa Conjunta n° 2, de 23 de Janeiro de 2006: Registro de Produtos Biológicos*. Ministério da Agricultura, Pecuária e Abastecimento, Brasilia.

MAPA (2006c) *Instrução Normativa Conjunta n° 3, de 10 de março de 2006: Registro de Produtos Microbiológicos*. Ministério da Agricultura, Pecuária e Abastecimento, Brasilia.

Meertens, B. (2008) *On-farms research manual (final draft)*. EU/CARIFORM Rice Project, Guyana–Suriname. Government of Guyana, European Commission and Agrifor Consult.

Meier,U. (2001) *BBCH Monograph: Growth Stages of Mono- and Dicotyledonous Plants* (2nd ed). Federal Biological Research Centre for Agriculture and Forestry (BBA), Berlin.

Pesticides Board Malaysia (2016) *Guidelines for Biopesticide Registration*. Pesticides Control Division, Department of Agriculture, Kuala Lumpur, p. 19.

SENASA (1999) *Manual de Procedimientos, Criterios y Alcances para el Registro de Productos Fitosanitarios en la República Argentina*. (Report No. Resolución Sagpya N° 350/99). Buenos Aires, Servicio Nacional de Sanidad y Calidad Agroalimentaria.

3   Sourcing Biopesticides

This chapter aims to support the process of sourcing biopesticides by leaf suppliers. It addresses some frequently asked questions and important considerations regarding the purchase of commercial biopesticide products. It also provides production guides for three biopesticides (*Trichogramma* wasps, neem and fungal biopesticides) for leaf suppliers who may be interested in exploring the local production.

3.1   Considerations for Sourcing Biopesticides

Topics covered in this section include:

- calculating the amount required per hectare
- identifying vendors
- time required for vendors to supply products
- storage

3.1.1   Calculating the amount required per hectare

To calculate how much of a biopesticide is needed in order to treat an area, follow these general steps:

3.1.1.1   *Step 1: How large is the area to be treated?*

- Multiply the length of the field by the width.
- Divide this number (in m^2) by 10,000 (the number of m^2 in a hectare). This is the number of hectares (ha).
- ha = length (m) × width (m) / 10,000 m^2/ha

    If feet are used, divide the number by 43,560 (the number of ft^2 in an acre). This gives the number of acres. Acres = length (ft) × width (ft)

3.1.1.2   *Step 2: How much biopesticide is needed to spray the area?*

- *What is the rate of application?*

It is important to read the label for the application requirement specified per hectare. Labels provide information on the biopesticide rate in metric measure such as litres per hectare (l/ha), millilitres per hectare (ml/ha), kilo-grams per hectare (kg/ha) or grams per hectare (g/ha).

- *If the measurement of field size is made in hectares,* then the biopesti-cide rate shown on the label can be used.
- *If the measurement of field size is made in acres (ac),* then it will be necessary to change the rate shown on the label from an amount per hectare to an amount per acre.
  This can be done by multiplying the rate by 0.4.

  - For example, 2l/ha × 0.4 = 0.8l/acre.
  - Amount per hectare × 0.4 = amount per acre.

- Once the rate of application has been calculated using the method de-scribed above, it is now possible to calculate the amount of biopesticide needed.
- Multiply the area to be sprayed (Step 1 above) by the rate of application.
- Amount of biopesticide needed = area to spray (ha) × biopesticide rate per hectare, or = area to spray (acres) × biopesticide rate per acre.

3.1.1.3   Step 3: How much area can be sprayed with one full tank?

Divide the tank size by the sprayer application rate.
The sprayer application rate is determined by calibrating the sprayer.

Tank size (l) = hectares that one tank will spray, or sprayer application rate (l/ha)
Tank size (l) = acres that one tank will spray, or sprayer application rate (l/ac)
Area sprayed (ha) = Tank size (l) / sprayer application rate (l/ha)

3.1.1.4   Step 4: How much biopesticide should be added to one full tank?

Multiply the area sprayed by one tank (Step 3 above) by the biopesticide rate being used.
Amount of biopesticide to add to a full tank = area sprayed by one tank (ha) × pesticide rate per hectare or area sprayed by one tank (ac) × pesti-cide rate per acre.

Note It is important to finish the complete tank mix in one spray. Storing the diluted biopesticide in the tank can lead to the loss of the efficacy of the ac-tive ingredient. Solution should never be stored in the tank.

3.1.2   Selecting among vendors

Channels for sourcing biopesticides may differ from that of conventional CPAs. Some examples of potential sources of biopesticides are as follows:

- **Distributors and agro-input retailers:** Some biocontrol manufacturers use existing local sales and distribution channels. This enables biocontrol manufacturers to penetrate the conventional market system. Local distributors have agents who receive advice from biocontrol manufacturers and then sell products to farmers. Biocontrol manufacturers will sometimes provide freezers to distributors (at sales points) for product storage and stocking. When buying from agro-input retailers, pay special attention to the expiry dates of biopesticide products as the movement of these products may be slower than pesticides usually supplied. In some countries, certain biopesticide products may be available for sale online from retailer websites, but this is not common.
- **Direct sales by biopesticides manufacturers:** Some biopesticide manufacturers have their own agents based in countries to sell and distribute directly to farmers. This route is more expensive as more in-country staff are required; however, it is a better way to build farmer trust and confidence through the provision of follow-up and support. Biopesticides manufacturers often collaborate to jointly market and distribute products and thus increase market access, e.g. Elephant Vert, e-nema, Andermatt Biocontrol, Real IPM (Biobest). In some countries, certain biopesticide products may be available for direct sale online from manufacturer websites, but this is also not common.
- **Universities and research institutes**: The research wings of universities and research institutes sometimes extend their research objectives into small manufacturing units attached to the labs to produce biopesticides. Though these are not widely advertised, some buyers such as farms near the facility might be aware of their existence and approach them for the product. The quality of such products is often more reliable compared with others, but availability may be an issue. To ensure sufficient supply of the biopesticides required, prior notice should be provided. Since production is often done at a low scale and without a commercial outlook, products produced at universities and research institutions may not have packaging for transportation and long-term storage.
- **Farmer organizations:** In order to better support crop production through their own resources, some farmer cooperatives and farmer-based organizations produce their own biopesticides, which they then provide to their members. Here both quality and availability of products can be an issue, as the products are often produced using low-tech methods and quality controls may not always be in place. Also, production may be carried out with the aim of immediate use and consequently product packaging may not be suitable for storage for any significant length of time.
- **NGOs:** Some non-governmental organizations working for the promotion of certain vulnerable sectors of society, such as women or rural youth, or those who wish to promote sustainable agriculture may produce and supply biopesticides.

When choosing among vendors, the following points should be taken into consideration:

- **Vendor knowledge of biopesticides:** Generally, national regulations will dictate that CPA vendors must possess a licence and also provide training in order to be able to sell CPAs such as biopesticides. As biopesticides are specific to the target pest, it is very important that the vendor has the right knowledge.
- **Turnaround time:** The shelf life for many biopesticide products is relatively short compared to that of conventional CPAs. Hence it is important to know that there will only be a short period of time between placing the order and procurement. Hence vendors located near such logistics services would be ideal.
- **Quality of the products:** The products supplied by vendors should be of standardized, uniform quality in order to avoid issues such as contamination or non-suitable substrates. Many countries have strict regulations defining standards and specifications for biopesticide products. Hence it is advisable to consider whether the biopesticide production units are audited by local authorities from time to time, drawing proper samples for quality check in referral laboratories and whether such records are documented. The product leaving the production units must comply with the quality assurance standard set by the company as a commitment to its customers. Although time-consuming, this will ensure the expected performance of the biopesticides in the field.
- **Range:** Since the best biopesticides are not broad-spectrum, the needs of farmers may be best met by vendors who have a wide range of products that can provide complete solutions to the farmers' issues. Hence vendors who have a variety of products for different types of pests could be a good resource for procurement.
- **Packaging and transportation**: With advancement in technology and better formulations the shelf life of many of the biopesticides has been enhanced, but transportation of such products is also important as they require fast and safe delivery. Packing the products with suitable material to avoid damage during transportation leading to loss of material is also important. Ensuring that packaging is as specified in any legislation is a must.
- **Availability:** Owing to certain interdependence on external factors, the production of biopesticide can sometimes be inconsistent. This may lead to products not being available at crucial times. Certain amounts of products should be stocked (bearing in mind expiry dates) to keep the supply consistent with demand. Some vendors have worked out the ways to make the supply consistent and would therefore be ideal to be flagged for procurement.
- **Facilities:** Stock should be kept under the right conditions to avoid any loss of the active ingredient due to unfavourable environmental conditions such as extreme temperature and humidity, which might lead to

secondary contamination. All biopesticide products should be kept in clean, dry and dark conditions that have slightly lower temperatures than the environment. Some of them, such as pheromones and nematodes, may also require refrigerated conditions. They need to be stocked on different shelves to those where chemicals are being stored. Vendors demonstrating such awareness of product stocking requirements would ideally be selected.

- **Order management:** Manufacturers of biopesticides use standard operating procedures and other approaches in order to keep their supply continuously consistent with the demands of customers while meeting a standard quality. For this, a robust system of order management is required, guiding processes right from the procurement for production until exit from the manufacturing unit.

3.2 *Trichogramma* Production Manual

Tobacco production globally is adversely affected by a number of pest insects, with lepidopterans playing a major role. Farmers often choose pesticides to control these pests, but increasingly there is pressure from the market to produce a crop that is free of hazardous residues and thus efforts are being made to use biopesticides for pest control. One effective agent that can be used to control lepidopteran pests of tobacco is *Trichogramma* wasps. These are very small (around 0.5 mm) parasitoids, which attack the eggs of lepidopteran hosts. Instead of pest larvae emerging from the eggs, more adult *Trichogramma* appear, which can then search for additional hosts in the crop. *Trichogramma* wasps are the most commonly used biocontrol agent in open field crops worldwide.

Trichogramma wasps can be mass reared, usually not on the target host but instead on a factitious (artificial) host that allows for easy production and maximizes cost, efficiency and output.

In the following manual, we aim to provide general guidance on the necessary steps, procedures and tools for the successful production of *Trichogramma*. The rearing of *Trichogramma* takes place in specialized facilities and involves two main steps: the production of the host, followed by the production of the *Trichogramma* wasps on the host eggs. A suitable factitious host on which most *Trichogramma* species can be reared is *Sitotroga cerealella*, also known by the common name 'Angoumois grain moth'. *Sitotroga cerealella* is a stored-product pest of grains and can be raised on a diet of good-quality barley and other grains (e.g. maize). This makes it easy to maintain in a laboratory setting for the mass production of *Trichogramma* (Fig. 3.1).

Even though the production of *Trichogramma* biocontrol agents is not particularly difficult, a practical training on the rearing methods is recommended before starting production.

Fig. 3.1 *Trichogramma* rearing facility in Laos (photo: Dirk Babendreier).

Topics covered in this section include:

- Outline of *Trichogramma* mass production procedure
- Basic rules for rearing
- *Sitotroga* rearing
- Rearing of the parasitoid *Trichogramma*
- Mass production of *Trichogramma*
- Egg cards, transport and releases
- Quality control

3.2.1   Outline of *Trichogramma* mass production procedure

3.2.1.1   Basic rules for rearing

- All personnel carrying out the rearing need to have received qualified training and must always follow the rearing operation protocol.
- Change shoes and put on a laboratory coat before starting work in the rearing rooms. Wash laboratory coats at least once a week.
- Keep the whole operation clean and tidy, and particularly free of organic material.
- The floor of the preparation room and the surface of the oven should be cleaned using a mop after sterilization.
- Kill/remove any larger animals, e.g. spiders, whenever they are found in the facility.
- Keep the environment surrounding the rearing rooms clean to minimize risks of infestations getting in from outside.

- Carefully plan the work that has to be done in order to avoid moving backwards and forwards between the rearing rooms too much.
- Take records on conditions (e.g. temperature), work done, and output; also note any particular observations.
- Develop and follow a well elaborated production plan, outlining details of how much needs to be produced and at which times, in order to meet demand from customers.
- Infestations with mites or other major pests (e.g. parasitoids attacking the *Sitotroga* larvae) in the rearing process may disrupt production and everything possible should be done to keep the risk of infestations low. To avoid infestations, note the following:

  o All the rooms should be sealed tightly from the outside environment; keep a second door between the rearing rooms and the environment and keep doors closed as much as possible.
  o Cut/remove any known sources of mites near the facility.
  o Sterilize the entire experimental *Trichogramma* rearing facility (ETRF) (including all equipment) before the start of *Trichogramma* rearing and at least once a year during the winter period; stick to necessary safety measures when doing so.
  o Clean the air fans within the production rooms from time to time and never use the air fans to draw air from outside, to prevent introducing pests from outside.
  o Try to minimize the number of entries and exits to and from the facility.

3.2.2   Mass rearing of *Sitotroga*

Sitotroga cerealella is a well-known storage pest with a global distribution. It is easy to rear and as such is commonly used as a host for rearing beneficial insects (Fig. 3.2). A founder population can be obtained commercially. As an alternative, it is also possible to start with a population collected from an infested local storage facility, but this method comes with the risk of beginning the rearing process with a contaminated starting culture.

As with nearly all lepidopterans, *Sitotroga* adults have scales on their wings that might affect the lungs of workers in the rearing facility. A small number of people also show a certain level of allergic reaction to the scales. Therefore it is important to keep the operation clean and to consider wearing masks in the rooms where adult moths are kept.

3.2.2.1   Diet preparation

- Clean, big-seed barley with high protein content is the best medium for rearing *Sitotroga cerealella*. About 10–15 g of host eggs can be harvested from 1 kg of barley.

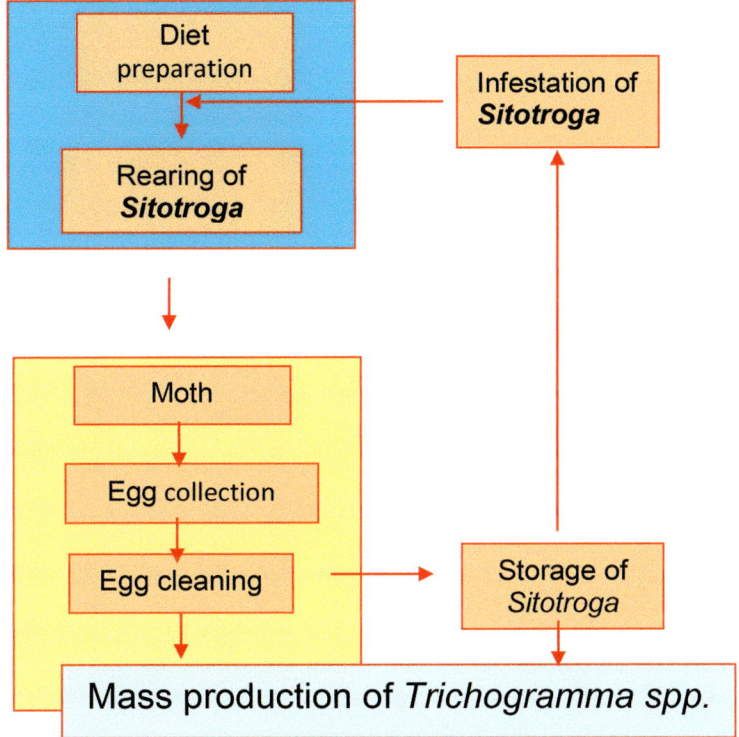

Fig. 3.2 Flow diagram showing general steps in *Sitotroga* rearing.

- Barley with loose glumes is preferred. Using barley with tight glumes will increase the mortality of newly hatched young larvae.
- The medium must be sterilized in order to avoid or minimize competition from other storage pests (e.g. *Tribolium* spp., *Sitophilus* spp). In addition, this step is essential in order to avoid infestations with competitive pests such as spider mites, predatory mites or even parasitoids. Infestations with the latter can severely disrupt production.
- Sterilizing the room with UV light will also help to prevent secondary infestations of the sterilized medium through fungi and other pests in the room.

3.2.2.2 Sterilization procedure

- Put the barley into water and stir it for 5–10 min; let the big healthy kernels sink to the bottom of the water container. Remove any damaged and small-sized kernels floating on top of the water.
- Transfer the barley into the sterilization container. The barley should be laid in the container at a depth of about 8–10 cm.
- Put the sterilization container into the drying oven straight away, together with any other material that will be used to handle the barley (e.g. a

stick for stirring). The barley should be sterilized at 80°C for 6 h. When it is ready, open the door of the oven, let the moisture out and let the hot barley cool down in the oven until the following day (do not use the hot barley immediately for host infestation).

3.2.2.3   *Infesting the medium with* Sitotroga

After the sterilization of the medium, it needs to be infested with eggs of *Sitotroga*. It is very important that the barley and the eggs used for infestation are clean and that the working place is clean as well. The number of eggs added should be adjusted to the amount of barley used.

- Boil some water and allow it to cool down before use; prepare clean, dry larval rearing containers.
- After sterilization, the barley needs to become moist again. Add 150 g boiled water for each 1 kg of barley; the barley is ready for use when it feels slightly damp, but not wet. If the barley is still too dry, add some more water using a hand-sprayer. Transfer the barley into the larval rearing container to a depth of around 3 cm (about 1.5 kg barley for each container).
- Prepare the *Sitotroga* eggs; check them under the microscope to make sure they are clean and free of pests (e.g. mites) (Fig. 3.3). Contaminated eggs should not be used so as to avoid heavy infestation during the host-rearing period a few weeks later.
- If the eggs are contaminated, try to obtain clean eggs from another source, e.g. commercially or from other rearing facilities. If this is not

Fig. 3.3 Checking host eggs under microscope for mites (photo: Dirk Babendreier).

possible, clean eggs can also be obtained 'internally' from the existing rearing facility, either by separating newly emerging moths from the moths that are obviously infested, or by treating the eggs:

- ○ **Separating moths:** Check carefully to see which cabinets are already infested with the pest. If uninfested rearing cabinets are present, then separate these from the infested ones. Try to use only those eggs obtained from the clean cabinets for further processing.
- ○ **Treating eggs:** In the case of larger pests, manual removal may be feasible but for mites, the only option is to treat eggs with an acaricide. To do this, dip the eggs into a 0.1–0.2% solution of acaricide for 10 min, then wash the eggs and let them dry before use.

- Measure 3–4 ml of *Sitotroga* eggs for each rearing container and carefully spread eggs evenly on top of the barley. Transfer the rearing container to the larval rearing room. The rearing room should be kept at a temperature in the range of 25–30°C, with 75–90% relative humidity (r.h.).
- It is very important to make sure that the eggs used are of high quality, which mostly means eggs that have been stored in appropriate conditions and for not too long a time.

3.2.2.4   Rearing of Sitotroga *larvae*

The larvae will hatch soon after infestation of the barley. They will be reared over the course of approximately 3 weeks, depending on the exact

Fig. 3.4 Larval rearing inside plastic trays (photo: Dirk Babendreier).

temperature they are kept at. During this time, they will feed on the barley grains (Fig. 3.4).

- Maintain the larval rearing room at 25–30°C and 75–90% r.h. Record the conditions in the rearing room daily. Be sure to maintain this high level of humidity so as to avoid high mortality of *Sitotroga* larvae. The room may be allowed to drop to 20°C, which will only result in slower larval development and production, but it must be kept lower than 30°C as temperatures any higher than this will cause the death of many larvae.
- Clean the rearing room regularly to avoid mite infestation.
- Switch on air fans for 1–2 h/day.
- On a weekly basis, check carefully that the rearing substrate is still damp enough, and spray (previously boiled) water on the barley if necessary.
- After 3 weeks of development, transfer the infested barley to the moth rearing room (see next step).
- Record infestation data: date of infestation / kg of medium used / ml of eggs used. This data will be needed later to calculate the average output of the rearing facility.

3.2.2.5   Rearing of Sitotroga moths

Rearing adults of *Sitotroga* is the last step needed in order to obtain eggs for parasitization and to keep the *Sitotroga* rearing going. At a temperature of 28°C, the moths will emerge after about 4 weeks. It is important to prevent infections by other pests such as predatory mites, spiders and other competitive pests.

- Clean and/or sterilize the moth rearing container and other materials. Transfer larvae-infested barley into the moth rearing container using the funnel.
- Transfer the moth collection container into the moth rearing cabinet.
- Keep the moth rearing room at 25–30°C, 75–85% r.h.
- Keep the rearing room absolutely clean.
- Switch on air fans for 1–2 h/day.
- Check for emergence daily and, after emergence has started, collect the moths 1–2 times/day.
- Check the rearing room conditions daily, make sure they are kept within the optimum ranges and take records.

3.2.2.6   Collection of Sitotroga moths

Collection of *Sitotroga* moths will be done once or twice a day, depending on the number of adults emerging. It should not be delayed because otherwise adults will start laying eggs in the moth collection cups. As much as possible, keep moths emerging on different days in separate cages. In general, moths may be kept in cages for 4 days after which time the cages can be emptied and cleaned. Moth emergence will usually continue for about 3 weeks.

Fig. 3.5 Moth collection containers placed inside the moth rearing cabinet, moth collection cup at bottom (photo: Dirk Babendreier).

- Collect the moths at least once a day; it is best to do this in the morning. At times of peak moth emergence, collect a second time in the afternoon. This is to avoid moths starting to lay eggs in the moth collection cup (Fig. 3.5). These eggs will be reddish in colour during the next day's collection time, meaning that they are of a reduced quality.
- Take a new, cleaned moth collection cup to the moth rearing room and exchange it for the old cup.
- Transfer the moths into the egg-laying cage and clean the cup.
- During times of high moth emergence, or when the moths are reluctant to move down as expected, an air compressor may be used to blow the moths down into the funnel. This method is only recommended when the facility is free of mites or other small pests which would be dispersed by the strong airflow.
- If the moths do not move down properly into the moth collection cup, check whether:

  - There is a high temperature gradient between floor and ceiling. This is usually caused by the use of the air conditioner.
  - The temperature is well above 28°C.
  - The r.h. is too low or too high (the optimum range is 75–85%).
  - There is an infestation by mites.

Fig. 3.6 Egg collection cage with eggs on tray (photo: Dirk Babendreier).

3.2.2.7   Sitotroga *egg collection*

Eggs will be collected from all the egg-laying cages every day, ideally more than once during peak production. Eggs may be collected on a piece of paper and stored in a cup or container before cleaning.

- The recommended density of moths within the egg-laying cage is 2–3/ cm^2. Overloading the egg-laying cage results in a reduction of oviposition because there is too much disturbance and the moths will become physically damaged while flying around the cage.
- For egg collection, place the egg-laying cage on the egg-collecting machinery (Fig. 3.6).
- Place a large piece of paper below the cage.
- Set the egg collection machine working for 5 min every 2 h for each of the cages in use.
- Collect the eggs once or twice a day by sweeping the eggs from the paper below the cage using a brush.
- Keep the room temperature at a preferred 24°C, 85% r.h. (high humidity increases egg laying by *Sitotroga* moth).
- Record egg collection details (output is recorded as ml of eggs).

Note

- If rearing is infested with other storage pests, check which moth-rearing cabinets are infested and keep the moths from these cabinets separate from the other egg-laying cages.

3.2.2.8   Cleaning of Sitotroga eggs

Only clean eggs should be used for the production of *Sitotroga* and *Trichogramma*. The collected eggs will be mixed with scales, legs and heads of the moths and these contaminants should be removed.

- The egg-cleaning funnel is used to clean away the moth scales. To do this, first connect the funnel to the vacuum cleaner.
- Place a tray under the funnel to collect the cleaned host eggs.
- Switch on the vacuum, and then slowly pour the newly collected host eggs into the upper opening of the funnel. Make sure the eggs are not clumped together as they fall down the airflow from the opening. Remove the light-weight scales and they will collect in the waste bag of vacuum cleaner.
- Make sure the power of the airflow is not too strong, so the eggs are not sucked out by the machine as well.
- Repeat the above procedure 3–4 times.
- This method will not remove the bigger debris such as heads and legs, which are too heavy to be sucked in by the vacuum cleaner. These can be removed by sieving the collected eggs with 40 and 60 mesh sieves.
- After cleaning the eggs, put them into a measuring cylinder to measure the quantity. Record the number of eggs on the data sheets.
- Put the eggs into glass or plastic tubs; seal the tubs with cotton cloths for storage (see details later).
- Moth scales are tiny and may clog the vacuum cleaner filter, so the filter must always be cleaned after use (disassemble and shake the filter part).

3.2.2.9   Storage of Sitotroga eggs

The correct storage of *Sitotroga* eggs is of utmost importance for successful and sustainable production. Host eggs will be stored for different purposes in *Trichogramma* production. If eggs are to be used for *Trichogramma* rearing, the eggs will be sterilized, killing the embryo inside, in order to avoid the emergence of larvae from eggs that are not parasitized. If the eggs are to be used for further rearing of the moth, the eggs must obviously be kept alive.

A carefully planned system is needed to optimize synchronization of host and *Trichogramma* production; storage of eggs is one component of this system.

- Eggs used for *Sitotroga* rearing should be kept at about 6°C (no higher than 10°C and no less than 2°C). They should be stored for no more than 10 days because longer storage time increases the mortality of larvae and adults, and also because moths produced from such low-quality eggs will have decreased egg production themselves.
- Eggs used for *Trichogramma* rearing should be kept between 10–15°C for the first 4–6 h, after which time the tube should be shaken to loosen the eggs in case they have stuck together. Thereafter the eggs should be stored at 3–5°C.
- For *Trichogramma* production it is best to use *Sitotroga* eggs that have been stored for no more than 1 week. Eggs stored for longer can also be used for *Trichogramma* production, but will be less attractive to *Trichogramma* adults, resulting in a reduced parasitism rate. The maximum storage time depends on the exact storage conditions and the quality of eggs in the first place, but may be around 2 months.
- Always make sure that the correct-sized containers are used in order to avoid egg desiccation. This means using well-sealed containers with not much air compared with the volume filled by the eggs.

3.2.2.10   Treatment of used medium

The used medium cannot simply be thrown away, as *Sitotroga* is a storage pest. Even after 3 months, there will still be a few moths emerging from the used barley, which could possibly infest storage facilities in the nearby surroundings, or wheat in the field.

- If no risk exists, i.e. there is no storage of grains nearby, then one of these recommendations should be followed:
  - Use the medium as a feed for pigs and chickens if it is dry.
  - Use it for composting or for producing organic fertilizer.

- If all the remaining *Sitotroga* need to be killed, either heat or cold treatments can be used. This might mean placing the material in an oven for about 2 h at 100°C, or in a freezer for a few days.

3.2.2.11   Sitotroga *culture in winter*

During winter time (or any other long out-of-season periods), the *Sitotroga* rearing should be maintained at a minimal level to enable production to be scaled-up when necessary in spring.

- To reduce workload and the number of generations, the temperature may be lowered as this will prolong the development time of the rearing host. However, temperatures should not fall below 10°C.

3.2.3 Rearing of the parasitoid *Trichogramma*

Trichogramma spp. are used on a larger scale in many parts of the world, in various crops. A species should be used that is known to be effective or likely to be effective against the targeted pests. It should also be commercially available in the region, or alternatively, a rearing colony can be started from collection of naturally occurring host egg masses in tobacco fields.

There are two phases in the rearing of the parasitoid: stock culture rearing and mass production. During mass production, the aim is to produce as many wasps for release as possible, based on the number of *Sitotroga* eggs. During the remaining times, the rearing is focused not on high output, but on the quality of the parasitoids reared, aiming to maintain a healthy and efficient stock for the next mass rearing cycle.

3.2.3.1 Sterilization of Sitotroga *eggs for stock culture and mass production*

In both these phases, eggs must be sterilized to prevent the development of embryos and hatching of *Sitotroga* larvae.

- To sterilize the eggs, thus killing the embryos inside the host eggs, spread the eggs in a single layer on an acrylic plate. If more than one layer is added, the lower layers will not be properly sterilized.
- Introduce the plate(s) into a cupboard about 30 cm below a clear UV-light tube (30W) and irradiate for about 20 min (Fig. 3.7). The precise time will depend on the exact distance to the light bulb and may need to be slightly adapted according to experience. Be careful to increase just

Fig. 3.7 Sterilizing host eggs under UV light (photo: Dirk Babendreier).

the time, because placing the eggs too close to the UV light will lead to desiccation, a major obstacle in *Trichogramma* production.

- Take care to protect eyes during the sterilization process by ensuring that the cupboard is not opened while sterilization is being carried out.

3.2.3.2   Stock culture

Stock culture is very important for maintaining the rearing cycle and ensuring the fecundity and efficiency of the *Trichogramma* wasps produced (Fig. 3.8). The whole process of stock culture rearing takes place in small containers such as glass tubes. To keep the stock culture healthy, introduce measures such as acclimatizing the *Trichogramma* wasps to outdoor conditions, adding different host eggs and/or introducing fresh parasitoids from the field.

- Keep the operation desk clean and spray with 75% ethanol if necessary.
- Egg stripes are prepared according to the following steps:
  - ○ Take sterilized, high quality *Sitotroga* eggs from the host rearing cage or fridge.
  - ○ Cut an adhesive tape to a length of several cm, stick a paper frame on the adhesive tape.
  - ○ Distribute the *Sitotroga* eggs evenly on the tape; they will stick to the middle of the stripe. Excess eggs can be removed using a fine brush and returned to storage.
- It is suggested that all stock rearing activities are carried out in glass tubes of about 12 cm length and 2 cm diameter. Egg stripes/cards should be placed inside the tubes according to the procedures described below. Seal the tube(s) using a fine screen or cloth and secure it with an elastic band. Put the glass tube into a larger, closed plastic or glass container containing saturated salt water. This system is used to control the relative humidity inside.
- To parasitize the egg stripes, take a card with eggs that have previously been parasitized by *Trichogramma* species from the existing stock. Check the status of *Trichogramma* emergence for this card according to the data record of parasitization. One to 2 days before the emergence of

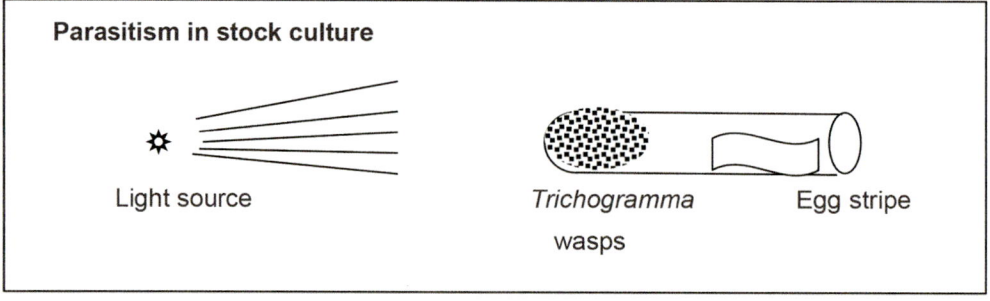

Parasitism in stock culture

Light source   *Trichogramma*   Egg stripe
                                wasps

Fig. 3.8 Scheme showing *Trichogramma* stock culture procedures.

the *Trichogramma* wasps, put some tiny dots of honey on the wall of the tube so that the first emerged wasps have food after emergence. N.B. The dots of honey must be extremely small otherwise the wasps will get trapped in the honey and die.

- During this critical period check at least twice a day, and as soon as the first wasps have emerged place a card containing unparasitized eggs in the tube. When opening the tube with active *Trichogramma* inside, always make sure that a light source is available and directed to the side of the tube, opposite to the opening, to attract the wasps and ensure that they do not escape from the tube. Seal the opening again with the screen/cloth and elastic band.
- Leave the tube in the laboratory while the *Trichogramma* parasitize the host eggs. If a large number of the *Trichogramma* have emerged on the same day in which fresh eggs were introduced, it is recommended that eggs are left to be parasitized for just 1 day; if emergence is less well synchronized then keep them in for 2 days.
- It is important to ensure that the ratio of the number of parasitoids to the number of new host eggs is kept at around 1:5 in the stock rearing. The *Trichogramma* from one egg stripe will be used to parasitize five new egg stripes. Most of the parasitoids from this one egg stripe will emerge during the course of 3 days. During these 3 days a total five new egg stripes will be exposed to the wasps. Usually, on day one only about 20% of *Trichogramma* will emerge and one new egg stripe can be introduced for parasitism. On days two and day three of the *Trichogramma* emergence period, two new stripes can be introduced per day.
- After 1 day of parasitism, remove the egg stripe (or the two egg stripes on days 2 and 3) again using a light source to attract *Trichogramma* wasps to the closed end of the tube. Put the newly parasitized eggs into a new tube, label it with the date of parasitism and keep them in the laboratory for development at 20–25° C.
- Make sure to carefully label the *Trichogramma* tubes used. Indicate when the parasitoids were used for the first day and also note what proportion of the parasitoids emerged.

3.2.3.3   *Degradation of the stock culture*

Problems can occur when insects are reared continuously in the laboratory for longer periods and over many generations. This can lead to parasitoids becoming less efficient or less healthy because they are adapted to the special conditions of the laboratory. When released, they might have difficulties with host finding, etc. In order to prevent this, certain practices can be adopted to keep the culture healthy. These are described below.

- **Adding fresh parasitoids** is the most efficient method. To do so, collect naturally parasitized eggs of the target hosts from tobacco fields and rear the *Trichogramma* from these. The best time for this is late season where there are usually more pests found and parasitism rates are higher, although there is of course a risk that *Trichogramma* species found may be different to that already being reared. To reduce this risk, it is best to collect in the centre of larger fields, and also consider consulting experts on species identification to avoid collecting the wrong species. This method is not recommended on a regular basis unless there is a clear need.

- **Adaptation *of Trichogramma* wasps to outside conditions**. Allow the *Trichogramma* wasps to parasitize the eggs at 20–25°C for 2 days, then take the tubes with the newly parasitized eggs outside and allow the *Trichogramma* to continue their development under outside conditions. After 5–6 days, take the tubes back to the labratory for the *Trichogramma* wasps to emerge. Follow this procedure over several generations.
- **Use target host eggs**. Another way to avoid degradation of the stock culture is to collect target moths in the field (e.g. using light traps). Take the moths to the labratory and rear them in a cage to collect eggs, and use these instead of *Sitotroga* for rearing the *Trichogramma* stock culture. When host eggs are substituted for at least one or even better for two to three generations, the degradation of the *Trichogramma* culture will be reduced.

3.2.3.4   *Mass Production of* Trichogramma

The mass production of *Trichogramma* is carried out in order to obtain enough material for large-scale releases (Fig. 3.9). From the stock culture, the culture is enlarged over and over until it is big enough for mass production. The procedures and equipment are different, although the basic steps for parasitism are the same as in the stock culture (sterilization of host eggs, parasitization with *Trichogramma*, storage of eggs). The last step is the production of egg cards, which is the final product for release.

- After sterilization, the eggs are parasitized by *Trichogramma* in the parasitization cage (Fig. 3.10). The attraction of parasitoid wasps to light is used in this cage as a way to keep them away from the cage door while adding fresh host eggs, and to attract them to the fresh eggs after introduction. The ratio of parasitoid female to host eggs is usually kept around 1:10 in order to produce a high number of parasitoids.
- Spread a thin layer of host eggs on to the acrylic plates and put as many plates as appropriate into one side of the *Trichogramma* parasitization cage.
- Shine a light on to the side of the cage containing the fresh host eggs.
- Put the *Trichogramma* stock into the mid-part of the cage, at the ratio of 10 eggs to 1 female wasp. Note that the sex ratio in the rearing cage will need to be known in order to be able to do this. Usually the sex ratio is between 50% and 75% females. Assuming that the ratio is 60% females and that 20 ml of eggs fit on one acrylic plate, about 3.3 ml of stock (parasitized eggs) will need to be added per plate. It is

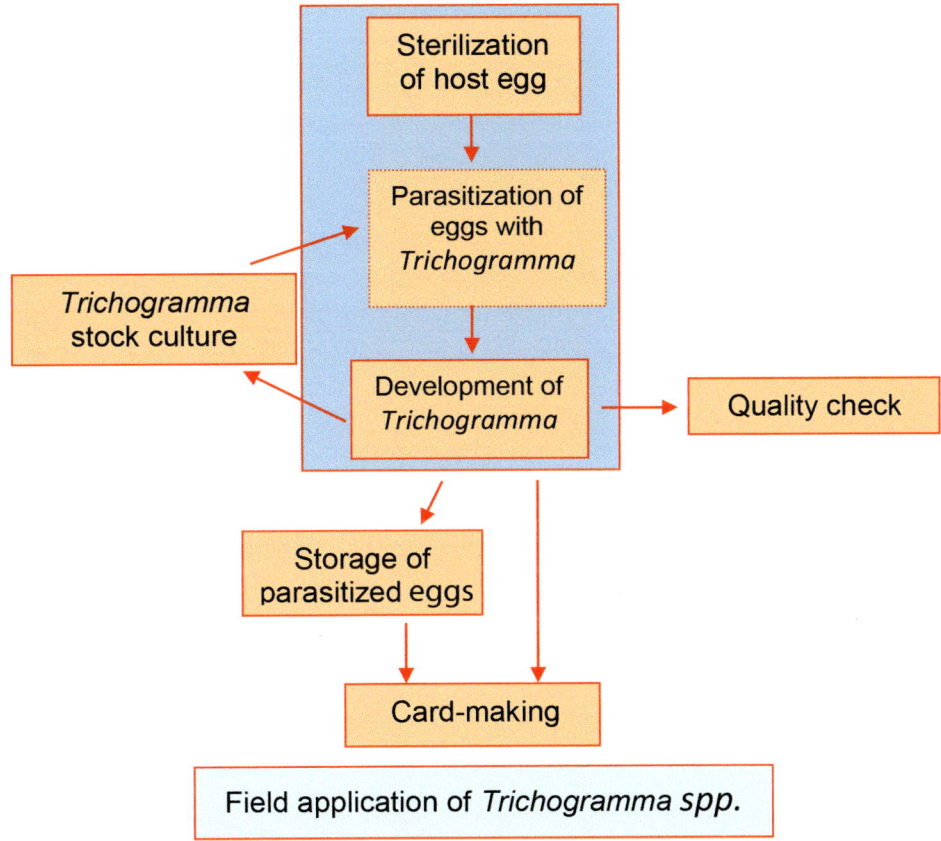

Fig. 3.9 Flow diagram showing general steps in *Trichogramma* rearing.

important that the stock material emerges over only a short period of time, and this can only be guaranteed if and when the correct stock culture procedures are followed (meaning that the exposure time of the wasps to the host eggs is only one day). Based on the parasitism rates achieved, the parasitoid:host ratio may be slightly adapted (Fig. 3.11).

- Repeat these steps 1 day later, i.e. put fresh host eggs onto new plates into the other side (which should still be empty) of the parasitization cage and add more wasps from the stock according to the number of host eggs introduced. Move the light to the side containing fresh host eggs.
- On the third day, remove the plate, which will have been in the parasitization cage for about 2 days, brush the eggs into vials and store at about 25°C for 2 to 3 days. New unparasitized eggs together with more *Trichogramma* stock can be added to the parasitization cage.
- The development time of *Trichogramma* depends on the temperature; lower temperatures prolong the development of the wasp (see table

Fig. 3.10 *Trichogramma* parasitization cage (photo: Dirk Babendreier).

Fig. 3.11 *Trichogramma* feeding on host eggs (photo: Gabriela Brändle).

below). This is important to take into consideration when mass rearing in order to allow for accurate timing of development for wasp releases.

Temperature	Time for Trichogramma development (egg to adult)
20°C	13 days
25°C	10 days
27–28°C	8 days
28–30°C	7 days

- About 2–3 days after removal from the parasitization cage (depending on room temperature), the parasitized host eggs will begin to turn to dark grey and then black. Once they have turned black, they can be easily checked for parasitization rate.
- At this stage, it is recommended to clean the eggs (by sieving) to separate the dead wasps and other debris from the parasitized eggs.

Notes

- The light on both sides of the parasitization cage should be of a low intensity.
- The humidity level should be kept high in the rearing room (75–85% r.h.), with a photoperiod of 14:10 light:dark. To avoid the risk of desiccation during the first days of storage the temperature should be kept under 25°C.
- The temperature should be around 25°C in the rearing room to ensure a high activity level of the *Trichogramma* wasps, but make sure it does not exceed 30°C.
- Add a small quantity of 10–20% honey solution in the central part of the parasitization cage using a piece of filter paper, but make sure the paper is not too wet, otherwise the wasps might get stuck.
- Keep an eye on the parasitism rate that is being achieved and adapt the wasp:host ratio accordingly, i.e. add more wasps from stock when the parasitism rate is lower than anticipated and vice versa. However, if the parasitism is still low when a high rate of wasps is added this may be an indication of problems with the quality of either the wasps or the host eggs (or both).

3.2.3.5 *Storage of parasitized eggs*

Developing *Trichogramma* wasps can only be stored for a limited period of time without seeing negative effects on their quality, i.e. loss of efficiency in the field or in further laboratory rearing. The storage conditions and rearing processes must therefore be monitored carefully to avoid problems.

- Measure the quantity of the parasitized eggs using a measuring cylinder.

- Leave the part which will be used without storage by putting the eggs in a tube and sealing the tube. Keep the eggs in the tube in the rearing room for further development until use.
- Put the eggs in a tube, seal the tube with a piece of cotton cloth and an elastic band and write the date on the tube. Fill in the data collection sheet.
- Transfer the tube containing parasitized eggs from the rearing room to an area with a lower temperature (15–20°C) for half a day. After this, transfer it to a fridge at 10°C.
- The eggs can be stored for 10 days before use. Eggs stored for 15–20 days can still be used after a check for their quality. If the eggs have been kept at a high r.h., and there is no sign of mould, they could still be used even after this time for parasitization, but there may be negative effects on their attractivity to adult *Trichogramma*.

Notes

- The most sensitive stages in the development of *Trichogramma* are the first day after parasitizing and the last day before emergence. Transferring such eggs from room temperature to a low temperature would cause high mortality of the *Trichogramma* and should therefore be avoided. The last day before emergence can be calculated using the table above (development time), but this requires that the rearing room temperature has been carefully maintained and recorded.
- At 10°C the development of the parasitoid larvae is drastically slowed, but not completely stopped, which means that the time they can be kept in storage is limited. If the eggs need to be stored for longer, they should be transferred to storage earlier (but within 1 day after parasitization).

3.2.4   Egg card production, transport and releases

The final step in the process is to use the eggs for egg card production and ultimately for release into the field. Several carrier substrates may be used for egg card production, the most common being paper/cardboard. The number of eggs per card will depend on the release system, mainly on the number of release points. As a rule of thumb, 100 release points are often sufficient for 1 ha of the crop. In this case 1000–1500 eggs would be placed on one egg card, resulting in a release rate of 100,000–150,000 parasitized eggs per hectare. Due to the variety of materials and designs that are possible for egg cards, only general steps are explained below, assuming the use of an egg card made of cardboard, with a sticky area of 1–2 cm^2.

3.2.4.1   Egg card production

- Before beginning to make the egg cards, loosen the eggs by sieving them (if necessary).

- Organize the egg cards with the sticky side facing up on a cleaned operation desk. Next, spread the parasitized eggs on to the sticky parts of the egg cards using a sieve.
- Sweep the eggs with a fine brush, to ensure that the eggs stick onto the tape.
- Clean the egg cards by carefully tapping the cards on the desk. Put the egg cards together, and note the production date.
- Remove any eggs remaining on the desk carefully using a brush; these can be used for further egg cards.

3.2.4.2   Transport of Trichogramma egg cards

- During transport, the parasitized eggs must be protected against heat, fumes of pesticides etc. A cool box can be used to provide an insulation layer. Place an ice pack or some other cold material inside to maintain a cool temperature, ideally in the range of 10–12°C. To ensure that the ice packs do not come into direct contact with egg cards they should be covered with materials such as crumpled newspaper or cotton.
- This method should allow the eggs to be kept cool for several hours. However, make sure that the parasitized eggs will reach their destination in the shortest time possible to avoid any damage to the egg cards during storage. Place a thermometer inside the transport box. If the temperature inside the box reaches over 35°C, the parasitoids will die without any changes being visible in the eggs.

3.2.4.3   Release of Trichogramma

- The best time to distribute the *Trichogramma* in the field is either early in the morning or late in the afternoon, since *Trichogramma* tends to emerge with the first light of the day. This timing also means that the *Trichogramma* are less likely to be lost due to the high daytime temperatures.
- It is best to distribute the *Trichogramma* just before wasp emergence. If the *Trichogramma* cards are placed in the crop several days before the emergence of wasps, there is a higher risk of the eggs being attacked by predators or damaged by climatic conditions (e.g. heat, rain). To minimize the risk of damage through solar radiation, the egg cards should be placed in the crop at lower levels, below leaves, or in the shade.
- Egg cards should not be released while it is raining.
- Releases must be timed according to pest phenology. For the best results sound monitoring schemes must be in place and information shared between *Trichogramma* producers, farmers and extension workers. Generally, the first release is recommended during the beginning of the flight period. Subsequent releases may be necessary later against the same generation and/or against a second generation.

3.2.5   Quality control

Quality control of *Trichogramma* rearing is important not only in terms of guaranteeing a supply of highly efficient biological control agents, which are needed to realize the anticipated increases in crop yield, but also to ensure the sustainability of the rearing. Quality control includes measurements taken during the production process in the facility as well as measurements taken in the field. The quality of the *Trichogramma* produced depends on many factors. The quality of the wasps must be checked regularly to make sure it remains high. If necessary, steps must be taken to improve wasp quality in cases where certain parameters have been negatively affected during the rearing process. Quality needs to be assessed for both the rearing host and the *Trichogramma* itself.

- Quality control at the facilities needs to be done several times during the production season. Always conduct quality control checks before planned releases.
- In addition to measurements taken at the facility, it is strongly recommended to collect data from the field as well, allowing relevant quality parameters of the released *Trichogramma* to be assessed.

3.2.5.1   *Adaptation by* Trichogramma *to the target pest*

In general, *Trichogramma* is reared on a factitious host such as *Sitotroga,* a convenient substitute for the real target pest, which cannot be reared in sufficient quantities or at a reasonable cost. However, after having been reared for many generations on *Sitotroga* eggs only, the parasitoid may lose its ability to parasitize the target pest(s). There are two ways in which this outcome can be avoided:

- Periodically collect parasitized eggs in the field and use these to refresh the culture.
- Maintain a rearing of the target pest and use its eggs for parasitization after a certain amount of time using the factitious host.

3.2.5.2   Trichogramma *quality control*

- Besides the adaptation to the target pest and keeping the correct *Trichogramma* species, the quality of the final *Trichogramma* product depends very much on the following (Fig. 3.12):
  - the number of eggs per card;
  - parasitism rate;
  - emergence rate: number of emerged parasitoids / total number of parasitized eggs;
  - sex ratio of the emerging parasitoids;
  - viability of the emerging parasitoids (measured via wing deformity);

Fig. 3.12 *Trichogramma* quality control (photo: Dirk Babendreier).

- In addition to the parameters mentioned above, activity of adult wasps (walking speed), longevity and lifetime fecundity are also good indicators of the quality of the *Trichogramma* but it is considered too time consuming to assess these parameters on a regular basis.
- The following items and materials are required to assess *Trichogramma* quality: graph paper, liquid glue (make sure it dries clear to avoid harming the insects), pins, paint brush, water, markers for labelling, counter (ideally with two categories), scissors, forceps, glass vials with screened lids, hand lens, microscope.
- **Technical procedures for *Trichogramma* quality control:** Cut 10 mm × 20 mm pieces of paper from the graph paper with scissors. Apply a thin layer of glue to one half of the 10 mm × 20 mm paper, then sprinkle about 100–200 eggs *Trichogramma*-parasitized eggs over the glue-covered area (be sure to handle the eggs gently). This number ensures that enough eggs will adhere to the paper to provide an adequate sample size, while at the same time minimizing the work. Between five and ten cards should be prepared for each control check.
- The 10 mm × 20 mm paper with eggs glued to it can then be placed into a labelled glass vial, which is closed with a lid and stored at about 25°C and 75% r.h. One day after the eggs turn black, the parasitism rate should be determined. It is also possible to do this at a later point, but it will be more difficult.

- After all the parasitoids have emerged, the quality control check can be conducted for all the other parameters. The *Trichogramma* adults that have emerged can be killed by placing the vials in a freezer for several minutes. If no freezer is available, the vials can simply be left at room temperature for a few days until the wasps are all dead.
- To assess the parasitism rate, place the egg card under a microscope and count all the black (parasitized) eggs as well as all the others. The parasitism rate should be higher than 70%, but excessive parasitism rates of close to 100% should be avoided.
- To assess the emergence rate, place egg cards under the microscope and look for emergence holes of *Trichogramma* wasps. Use a pin to remove the eggs that have been counted to avoid recounting them. For eggs without evident holes, use the pin to rotate or open them to look for signs of parasitism. In some cases, remainders of the parasitoid or even fully-grown adults that did not manage to emerge will be found. If any liquid material erupts from the eggs, this means that the *Trichogramma* adults have not successfully emerged.
- Parasitized eggs of good quality show a high emergence rate (percentage of black eggs with holes) of between 90 and 100%. If the emergence rate is lower than 85%, check whether:

  o The eggs are over-parasitized (too many parasitoids have developed inside one host egg). Check the abundance of *Trichogramma* parasitoids in the parasitization cage and reduce the number if necessary.
  o The parasitized eggs have been stored under the wrong conditions (temperature lower than 10°C or higher than 35°C), the eggs have been stored for too long in the fridge at 10° C (more than 10 days), or the eggs were stored close to pesticides and other chemicals.

- The sex ratio can be assessed by placing all the emerged *Trichogramma* adults under the microscope. Looking at one individual at a time, determine the sex (this is best done by looking at the antennae, which in the males have many hairs) and then use a wet paint brush to remove it from the collection to avoid recounting it. The sex ratio of the emerged parasitoids should be more than 50% females. A high ratio in favour of males may indicate overcrowding in the parasitization cage.
- Wing deformity can be assessed at the same time as checking the sex ratio, again looking at one individual at a time. Determine if the wings of each adult are deformed or not. The number of individuals with undeveloped wings should be very low (less than 5%).

3.2.5.3   Sitotroga *quality control*

- The most important quality criteria for *Sitotroga* is the hatch rate (proportion of healthy host larvae hatching from the eggs).

- Another relevant parameter is egg size, since *Sitotroga* eggs that are too small will not support the development of a *Trichogramma* adult.
- Using the same materials and procedures as above for *Trichogramma*, small (10 mm × 20 mm) pieces of paper need to be prepared on which about 100 freshly laid (and thus unparasitized) eggs are sprinkled. These pieces of paper are placed in a labelled glass vial, sealed with a lid and stored at about 25°C and 70–80% r.h. After 5 days, check how many larvae have hatched. For each quality control assessment, 5–10 samples should be checked.
- The normal or 'perfect' hatch rate is over 95%, but slightly lower rates can be acceptable. If hatch rate becomes less than 70%, something must be wrong and action needs to be taken.
- The most likely causes for low hatch rate are high temperature or low humidity, but it could also be due to bad-quality grains being used.

3.2.5.4 Quality control measurements from the field

In order to assess the quality of *Trichogramma* in the field, some of the egg cards that have been placed in the field for release should be recollected to measure the parasitism and emergence rate.

- Egg card recollection: to facilitate the recollection of egg cards from the field for quality control, the plants on which cards have been placed should be marked. Ten cards should be marked and recollected as soon as *Trichogramma* emergence has ceased (typically 2–4 days after placement in the field). These cards must then be analysed for emergence rate (see above). Other potential problems such as predation on eggs (any eggs that have been eaten or are missing), or egg cards that have fallen to the ground, should also be noted.
- In addition, information on parasitism rate is very useful to have in order to assess the success of the release, and indirectly the quality of the material released. However, this requires more field work on both release plots and control plots, and is beyond the scope of this manual.

3.3 Neem Production Manual

Azadirachtin is a naturally occurring substance found in the neem tree (*Azadirachta indica*), which can be used to prepare homemade or even commercial insecticides. Azadirachtin can act as an insecticide, an insect repellent, an anti-feedant and an oviposition deterrent. It is used against a wide range of pests such as aphids, whiteflies, thrips and flea beetles. Neem-based products (clarified hydrophobic extract of neem oil) have also been demonstrated to have some fungicidal properties, acting to reduce growth of fungus and the formation of fungal spores, which spread the disease. Neem also forms a film on the leaves that prevents the fungus from settling.

Neem products can be derived from the leaves, leaf extract, seeds, cakes, oil or fruit extracts; however, azadirachtin is most concentrated in the seeds. The products can be used as insecticides to prevent and treat seed and soil pests and other pests of tobacco. The concentration of the active ingredient (azadirachtin) found in the product may vary depending on the methodology used to extract it.

Topics covered in this section include:

- General recommendations for the preparation of neem solutions
- Neem seed extract
- Neem leaf extract
- Neem oil spray
- Neem powdered seed extract.

3.3.1   General recommendations

When controlling pests using neem-based products prepared at home, the following are the standard procedures for their preparation and application:

- Select plants/plant parts that are pest-free.
- When storing the plants/plant parts for future usage, make sure that they are properly dried and stored in an airy container (never use plastic containers), away from direct sunlight and moisture. Make sure that they are free from mould before using them.
- Make sure that the utensils used to prepare the extract are not used for food preparation or for drinking- and cooking-water containers. Clean all the utensils thoroughly after every use.
- Do not come into direct contact with the crude extract during preparation or application.
- Make sure that the plant extract is kept out of reach of children and house pets while left overnight.
- Always test the plant extract formulation on a few infested plants first before moving on to large-scale spraying.
- Wear protective clothing while applying the extract.
- Wash hands after handling the plant extract.
- Generally, oil is added to the solution in order to help it stick to the plant and soap is added to disperse the oil.

3.3.2   Neem seed extract

Two methods can be used:

Method of preparation (1)

- Collect mature, ripened fruits, remove the flesh and dry the seeds.
- Gently pound 3–5 kg of de-shelled neem seeds.

- Place the pounded seeds in a clay pot. Add 10 l of water.
- Cover the mouth of the pot securely with a cloth and leave it for 3 days. Strain to obtain a clear extract.
- Dilute 1 l of neem seed extract with 9 l of water.
- Add 100 ml of soap or vegetable oil. Stir well and spray.

Pests controlled

- Most agricultural pests, including, aphids, thrips, whiteflies, flea beetles, Lepidoptera, grasshoppers.

Method of preparation (2)

- Add 50 g of powdered kernel to 1 l of water.
- Let it stand for 6 h but no more than 16 h.
- Add soap and stir. Constantly shake the container or stir the extract during the process of application.

Pests controlled

- Aphids, American bollworms, grasshoppers, whiteflies.

3.3.3 Neem leaf extract

Method of preparation

- Collect and chop fresh neem leaves.
- Gently pound 1–2 kg of neem leaves.
- Place the leaves in a pot and add 2–4 l of water.
- Cover the mouth of the pot securely with the cloth and leave it for 3 days.
- Strain to obtain a clear extract. Dilute 1 l of neem leaf extract with 9 l of water.
- Add 100 ml of soap and stir well.

Pests controlled

- Aphids, Colorado potato beetles, grasshoppers, grubs, Japanese beetles, leafhoppers, locusts, plant hoppers, snails, thrips, weevils, whiteflies.

3.3.4 Neem oil spray

Method of preparation

- Add 30 ml of neem oil into 1 l of soapy water.
- Constantly shake the container or stir the mixture during application to prevent oil from separating.

Pests controlled

- Flea beetles and soft-bodied insects like aphids, mealybugs, mites, thrips and whiteflies.

3.4   Manual for the Mass Production of Fungal Biopesticides

The production manual presented in this chapter aims to make microbial biocontrol agents more accessible. The focus of this manual is the mass production of fungal biocontrol agents, and the guidelines provided focus on semi-solid or two-phase mass production of fungal biocontrol agents. Other technologies also exist, in particular, large-scale production in bioreactors (liquid fermentation). The methods proposed here are more amenable to adaptation to a range of situations from semi-commercial production, laboratory-scale or even low-technology production in less developed countries, even at the community or farm level.

The mass production of a number of well-known fungal biocontrol agents follow general approaches and principles, which are presented here. However, the main methodologies presented assume some prior knowledge or the availability of technical support for basic microbiology techniques such as aseptic or sterile techniques. In addition, due to variations among species and strains/isolates and growth requirements/rates some adaptation of the production systems will often be necessary for optimal production.

The main groups covered by this manual are *Beauveria, Metarhizium* and *Trichoderma*. Where applicable, differences in production methods will be highlighted and any modifications to the production methods required for more low-technology systems will also be indicated.

This manual covers all the steps for mass production of fungal biopesticides (Lomer and Lomer, 1998), including:

- Mass production: semi-solid substrate or two-phase
- Preparation of fungal spore inoculum
- Preparation of liquid starter culture
- Preparation and inoculation of solid substrate
- Harvesting and drying of spores
- Quality control
- Non-routine procedures
- Annexes
- References

3.4.1 Mass production: semi-solid substrate or two-phase

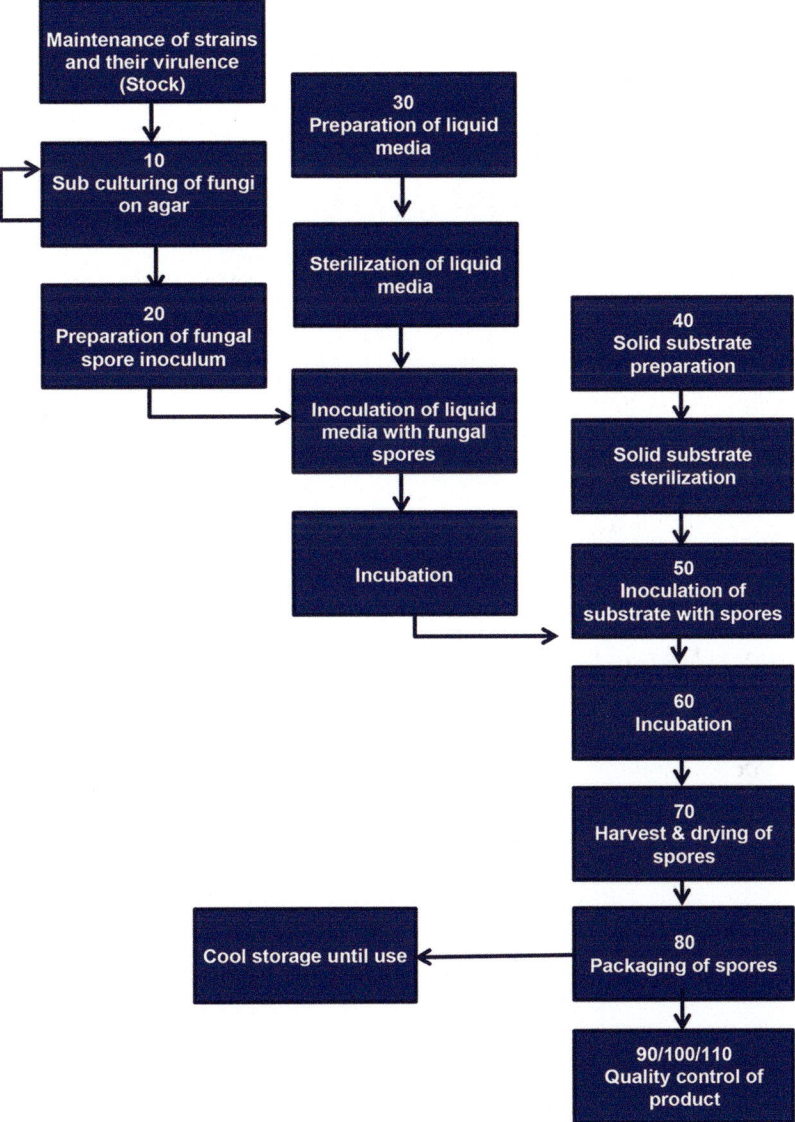

Fig. 3.13 Overview of two-stage mass production process.

3.4.2 Preparation of fungal spore inoculum

The fungal agents dealt with in this manual all have similar growth requirements in culture, so a general methodology will be presented, with any

variations being highlighted (Fig. 3.14, Fig. 3.15). Key organisms to be produced using this manual are:

- *Beauveria bassiana*
- *Metarhizium* species (e.g. *M. anisopliae*)
- *Trichoderma* species (e.g. *T. harzianum*)

Further information on these organisms can be found in Chapter 4.

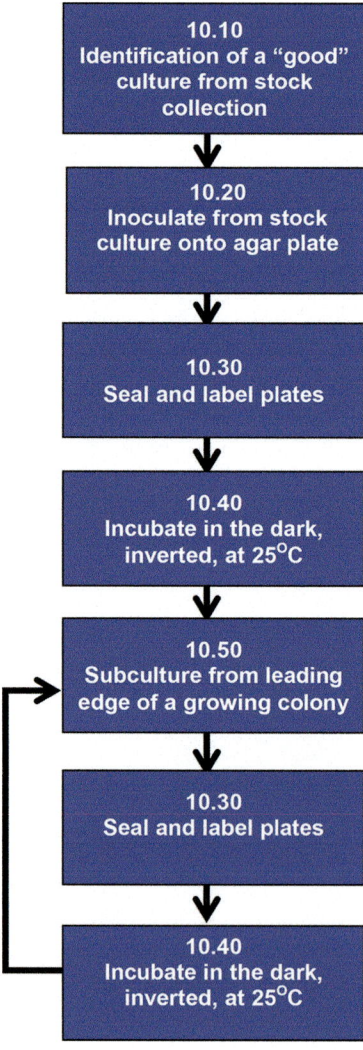

Fig. 3.14 Flow diagram for subculture of working cultures.

All procedures should be carried out using standard aseptic techniques. Unless otherwise stated all actions should be carried out close to a naked flame (Bunsen burner or alcohol burner), and if available use a laminar flow cabinet.

Flow chart	Instructions	Important points/tips
10.10	• Select the stock culture to subculture fungus for the working cultures. Ensure it is clean and free from contamination.	• Stock culture may be maintained as an agar plug in water, under mineral oil or as a cryopreserved sample (see Annex 3.4.8.2). It may require time to produce a useable colony.
10.20	• Inoculate onto 9 cm agar plate(s). Remove agar plug or small amount of mycelium or spores, dependent on the source material, and place in the centre of the agar plate.	• A range of agars may be used for the subculture, potato dextrose agar (PDA), 20% PDA and potato carrot agar (PCA). 20% PDA or PCA are preferred for sporulation, full PDA often promotes more vegetative growth. Saboraud dextrose agar (SDA) can be used for entomopathogens (see Annex 3.4.8.1 for media recipes and preparation).
10.30	• Seal plates with parafilm and clearly label.	• Always label the bottom of the Petri dish, never the lid. Use a waterproof marker.
10.40	• Incubate the plates in an incubator at 25°C, in the dark, or under a near-UV light (black light).	• Invert petri dishes, temperature of incubation may vary depending on fungal isolate.
10.50	• Remove an agar plug from the leading edge of a growing colony, using a 5 mm core borer and place centrally on the agar. • Prepare as many plates as required for further subculture, or for inoculum production.	• Observe colony growth regularly to ensure correct timing of subculture. The growth time may vary for species and/or isolates. For slow growing fungi a spread plate can be made with spore solution. • If required check culture under the microscope (see Chapter 4).

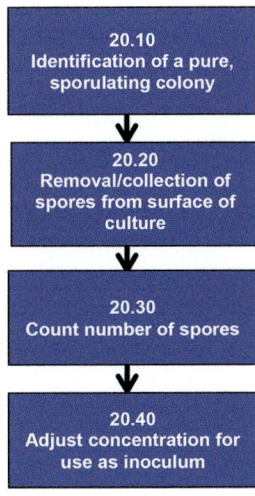

Fig. 3.15 Flow diagram for preparation of fungal spore inoculum.

Flow chart	Instructions	Important points/tips
20.10	• Before harvesting spores ensure the colony (or colonies) to be harvested is sporulating. Ensure colony is free from contaminants.	• Visual inspection should be sufficient. • 10-day-old colonies should ordinarily have sufficient sporulation.
20.20	• Flood the colony with sterile distilled water (SDW) containing 0.05% Tween 80. • Use a glass rod, or sterile rod, to gently remove spores from the surface of the colony. Use a sterile syringe to collect the spores and transfer to a sterile glass universal bottle.	• Where distilled water is not available sterile tap water may be used. Similarly if Tween 80 is unavailable then another detergent may be used. • If spore solution contains substantial mycelia or other debris then filter through sterile lens tissue (or cheesecloth/muslin), to collect spores and put into a new sterile universal glass bottle.
20.30	• Use a haemocytometer to count the number of spores (see Annex 3.4.8.3).	
20.40	• Adjust the concentration to the required number of spores per ml using SDW with 0.05% Tween 80.	• Spore concentrations of 1 to 6×10^6 spores/ml are typically used for inoculum.
		• Viability of spores can be assessed by removing 0.1 ml of suspension and using a glass spreader to distribute them over the surface of a PDA plate. • After 16–24 h percentage germination can be assessed under the microscope. • Spores are considered to have germinated if the germ tube length is half the spore diameter (see Chapter 4).

3.4.3 Preparation of a liquid starter culture

A liquid starter culture can be prepared to add as an inoculum to the solid substrate (Fig. 3.16). This can increase the rate of colonization and reduce the likelihood of contamination in the production system. Alternatively, spore inoculum, as produced above, could be used directly to inoculate the solid substrate.

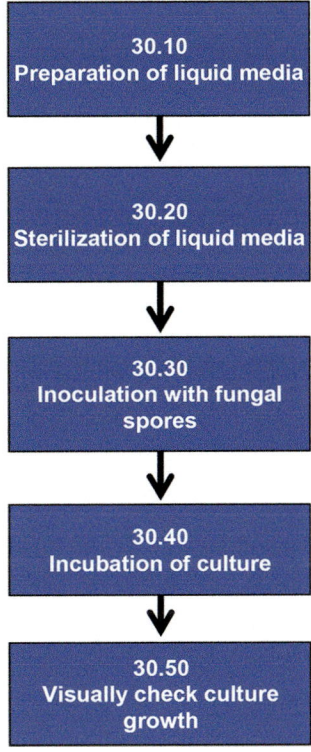

Fig. 3.16 Flow diagram for preparation of a liquid starter culture.

Flow chart	Instructions	Important points/tips

30.10	A range of liquid media may be used, including potato dextrose broth (PDB), molasses yeast extract medium (MYE), basal liquid medium (BLM) or brewer's yeast/sucrose broth (BYSB) (see Annex 3.4.8.1 for media). Here we will use brewer's yeast/sucrose broth as an example;
	• Weigh 10 g yeast and 10 g sucrose, using a balance, into a 1 l flask/bottle and add 500 ml tap water.
	• Boil to dissolve the yeast and sucrose (microwave for 7–9 min).
	• Decant the medium into 75 ml aliquots in a 250 ml conical flask.
	• Into the opening of each flask insert a breathable sponge bung, or non-absorbent cotton wool, and cover with tinfoil.
		• Assess initially which medium is best for organism to be used:
		  ○ PDB, MYE, BLM: *Trichoderma* spp.
		  ○ BYSB: *Beauveria* and *Metarhizium* spp.
		• Volumes may vary depending on the quantity of inoculum.
		• If no rotary shaker is available use glass medical flats, for static culture (see 30.40). A thin layer of medium allows greater gas exchange.

Flow chart	Instructions	Important points/tips
30.20	• Autoclave the flasks for 15 min at 121°C/15 psi. • Leave to cool before further use. • Contamination check: pour a small quantity of the liquid medium onto a dish of solid medium, seal dish with Parafilm™, put at 25°C in incubator and then check for contaminant growth after 48 h.	• If no autoclave available use a household pressure cooker, 30–45 min. • Use autoclave tape for each sterilization batch to ensure effective autoclaving has occurred. • Autoclave period may be extended to 20–30 min. • Ensure careful release of sterilizer pressure (cotton plug must stay dry).
30.30	• Aseptically, remove the tinfoil and bung from the flask opening. • Aseptically add 1 ml of spore inoculum to each flask of liquid medium using sterile pipette tips. Replace all bungs.	• Spore inoculum can be 1 to 6×10^6 spores/ml. • Flame after opening and before closing. • Take some of the remaining spore inoculum and check for contamination (see Chapter 4).
30.40	• Incubate conical flasks on a rotary shaker at 150–175 rpm, 25°C for 3 days or as required from experience.	• For medical flats incubate until a mat has formed on the surface of the liquid and check for sporulation. • For *Trichoderma*, 4 days on BLM medium can be used, or 5 days with MYE.
30.50	Visually check flasks for contamination.	• If it appears cloudy from bacteria or yeast then discard.
	Inoculum can be used directly, diluted or initially blended (if static culture). If standardization of the spore inoculum is needed, then filter through muslin or cheesecloth (aseptically), as required.	

3.4.4 Preparation and inoculation of solid substrate

The solid substrate stage of the production process provides a surface area for the fungi to produce aerial conidia. A large surface area per volume is preferred. White rice is often the material of choice (Fig. 3.18) but other cereal or cereal by-products may be used.

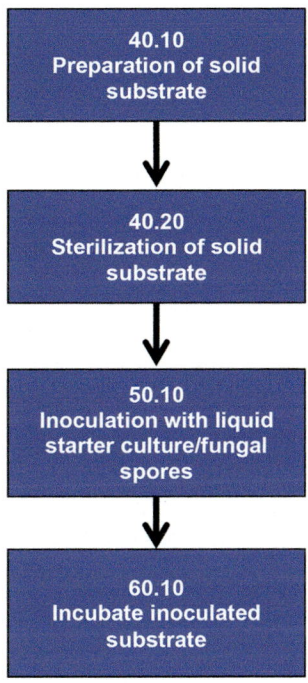

Fig. 3.17 Flow diagram for preparation and inoculation of solid substrate.

Flow chart	Instructions	Important points/tips
40.10	A range of solid substrates can be used, the most commonly used material being rice. A method for the preparation of rice as a solid substrate is provided here: • Place 1 kg rice into an automatic rice cooker. Stir in 300 ml tap water and cook. • Roughly divide the cooked rice into quarters and put each into an autoclave bag. • (Repeat until enough rice for the production volume has been cooked.) • Take one large autoclave bag and place a strip of autoclave tape on the inside. Use this bag to hold six bags (neck of the bag is folded over, but not sealed) of cooked rice. Alternative method, no pre-cooking of rice: • 300 g of rice soaked in tap water for 40 min. • Non-absorbed water is drained and the rice placed in polypropylene bags. • Bags are filled so that there is still space for aeration and mixing of substrate. • The neck of the bag is folded twice to ensure aeration but maintain sterility once removed from autoclave (Krauss *et al.*, 2002).	• Adjust volumes as required for production. • If the rice contains a lot of starch, thoroughly wash it under a running tap before cooking. • If the rice is particularly sticky, stir 2.5–10 ml vegetable oil into the rice prior to cooking. • Depending on the isolate, the water content may need adjusting. • Plastic autoclavable bags may be used if no autoclave bags available. • Alternatively, plastic bags with filters for gas exchange may be used, if available. • If a rice cooker is not available follow alternative method with no pre-cooking.

Flow chart	Instructions	Important points/tips
40.20	<ul><li>Autoclave all the bags containing the cooked rice for 40 min at 121°C/15 psi using an autoclave.</li><li>Leave to cool.</li><li>On removal from the autoclave make sure that the bag tops are kept firmly closed to prevent contamination.</li></ul>	<ul><li>To ensure all material is sterilized, some methods advocate sterilizing the bags on two or three consecutive days.</li><li>Bags are weighed before and after third autoclaving to calculate water lost. This is then added back with the liquid starter culture/inoculum. If available, place autoclave tape in bag before autoclaving, to confirm sterility.</li></ul>
50.10	<ul><li>Take flasks of fungal starter culture produced in step 30.</li><li>Aseptically, add a minimum of 10 ml of inoculum to each 100 g of substrate, e.g. 50 ml to 500 g of rice. For ease add contents of flask, approximately 75 ml.</li><li>Fold the neck of the bag over and fasten at three points with staples; leave loose to allow aeration.</li><li>Massage the rice in the bag to mix.</li></ul>	<ul><li>Alternatively, spore inoculum may be used directly from agar plates, though there will be a lower colonization rate.</li><li>May also apply specific colony-forming units (c.f.u.)(see Chapter 4).</li></ul>
60.10	<ul><li>Transfer the bags of rice to shelving in clean room. Spread the rice within the bag and tent the top so that the bag does not touch the top of the rice.</li><li>Leave at room temperature for 7–14 days, until conidiation is observed.</li><li>Shake bags every 3 days to improve aeration.</li></ul>	

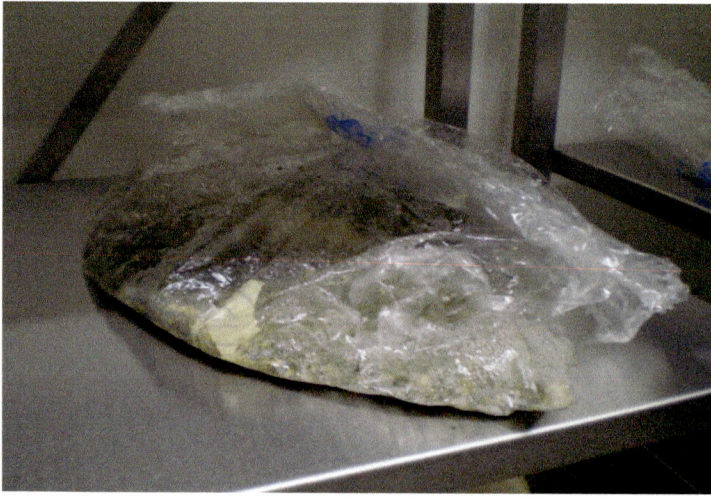

Fig. 3.18 Fungal spores growing on rice in a plastic bag (photo: E. Thompson, CABI).

3.4.5 Harvesting and drying of spores

The spore-covered substrate may be used directly and incorporated into the soil or washed to remove spores, which are then applied as an aqueous or oil solution (spray or drench). However, it is often preferable to harvest the spores and store until use. Drying the spores to below 5% moisture content will enable more long-term storage and maintain the viability of the spores (Fig. 3.19).

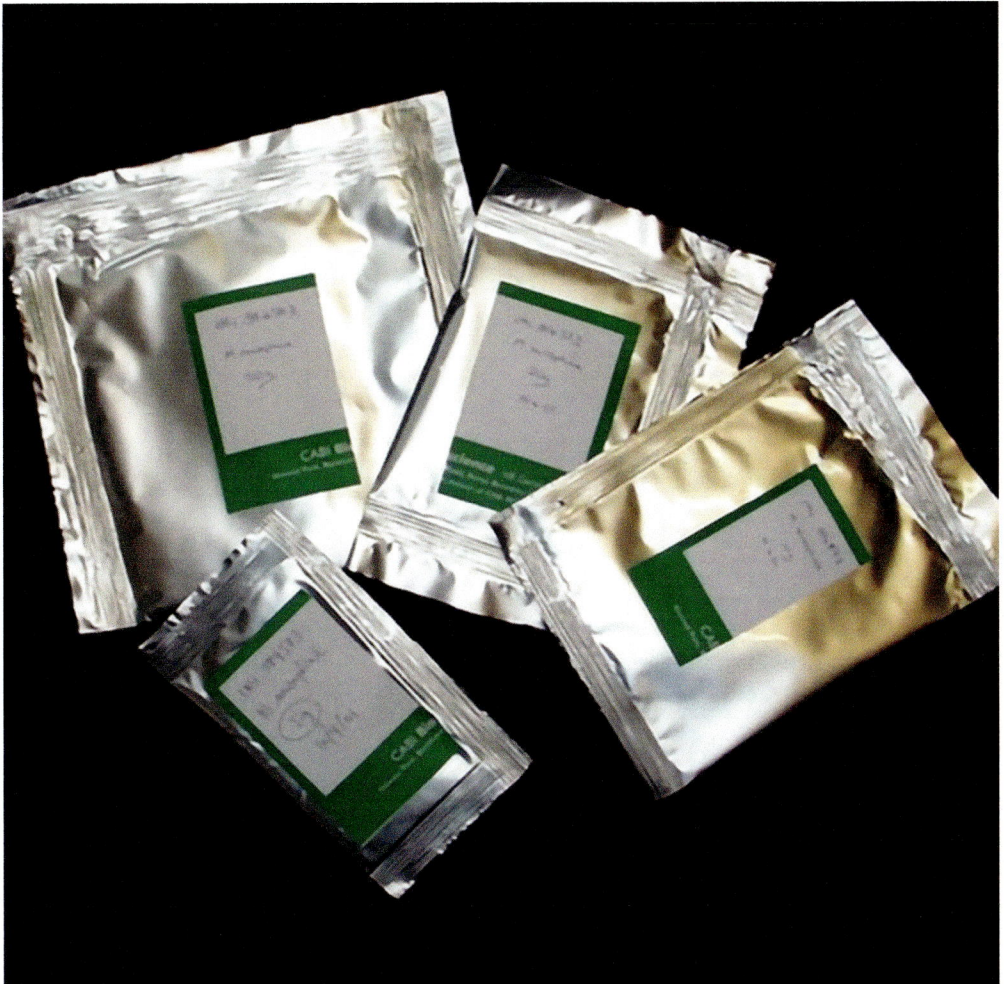

Fig. 3.19. Tri-laminate sachets of dried spores (photo: E. Thompson, CABI).

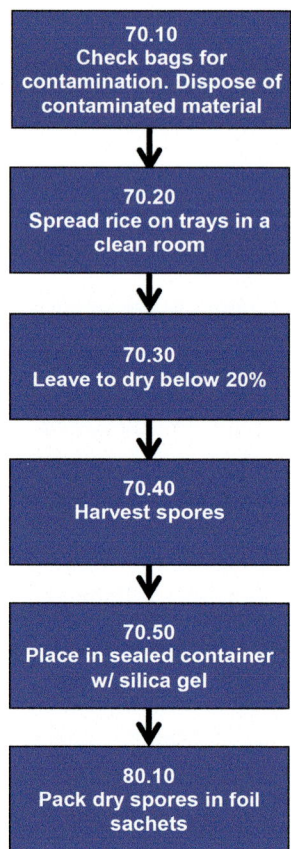

Fig. 3.20 Flow diagram for the drying of spores.

Flow chart	Instructions	Important points/tips
70.10	• Check bags of solid substrate from 60.10 for contaminants. If any evidence of contaminant dispose of bags.	
70.20	• In a clean room place contents of bags onto foil lined trays. • Spread out the rice and crumble any clumps.	• Room should ideally be at 20–25°C with 50–55% relative humidity.

Flow chart	Instructions	Important points/tips
70.30	• Leave to air dry until the moisture content of the rice drops below 20%, using a moisture analyser to measure the moisture content. • When the moisture content of the drying rice is just below 20% the conidia are ready to be harvested.	• If high moisture content in the atmosphere, use a dehumidifier in the room before placing rice in it. • If no moisture analyser available, weigh sample at the start and subsamples subsequently, to assess moisture content. (Oven dry a weighed subsample at 105°C for 24 h, weigh, then assess water content.)
70.40	• Harvest conidia from the rice either using MycoHarvester or sieves. If using the latter, firstly sieve with 300 µm, then 105 µm and finally a 75 µm to harvest conidia suitable for an ultra-low volume sprayer.	• NB. When harvesting, ensure respiratory protection is worn (e.g. dust mask or respirator). • Wear gloves if handling spores.
70.50	• After harvesting, place the conidia inside a closed container on top of dried non-indicating silica gel beads. • Regularly refresh the beads by drying in an oven for 2 h at 70°C and leaving them to cool before use. • Repeat until the moisture content is just below 5%, using moisture analyser to measure moisture content.	• Confirm identification of final product (see Chapter 4). • Assess viability of spores and number per gram (see Chapter 4 and Annex 3.4.8.3).
80.10	• Package in sachets (tri-laminate foil) and store at around 5°C. • Label all sachets with the mass production run number, fungal species and date of packaging.	

3.4.6 Quality control

Quality control can be considered to encompass two broad areas: process quality control and product quality control.

3.4.6.1 *Process quality Control*

Process quality control covers the confirmation of the purity of the product during production and ensures that contamination is prevented.

A key point is to be able to clearly identify the fungal material to be mass produced and differentiate it from contaminants.

This manual is focused on the production of three main fungal genera:

Beauveria spp.
Metarhizium spp.
Trichoderma spp.

Main characteristics of each group:

Beauveria spp.

Colony morphology

On PDA or malt agar, colonies are velvety to powdery, white at the edge becoming pale yellow to pink/red. The underside of the colony is colourless, yellowish or reddish (Brady, 1979a).

See Fig. 3.21, Fig. 3.22 for microscopic views.

Key reference for identification:

Brady (1979a) CMI Descriptions of Pathogenic Fungi and Bacteria No. 602, *Beauveria bassiana*.

Metarhizium spp.

Colony morphology

On PDA after 14 days colonies have a white mycelial margin with clumps of co-nidiophores, which become coloured with the development of the spores. The colour varies from olivaceous buff to yellow-green or olivaceous to dark green, but sometimes pink or vinaceous buff. The underside of the colony is colour-less or honey buff, sometimes with a yellow pigment diffusing into the medium.
Microscopic view is shown in Fig. 3.23.

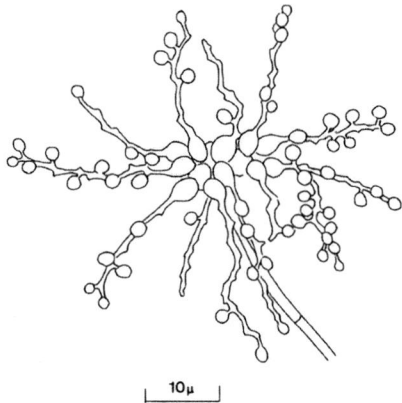

Fig. 3.21 Conidiophore and spores of *Beauveria bassiana* (CMI Descriptions of Pathogenic Fungi and Bacteria No. 602).

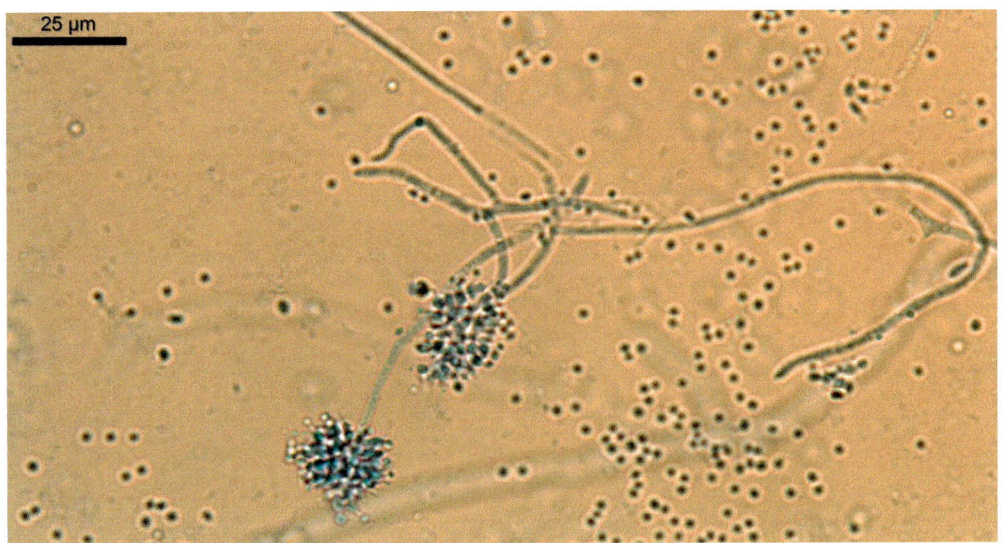

Fig. 3.22 Spores of *Beauveria bassiana* on conidiophores (photo: Sarah Thomas, CABI).

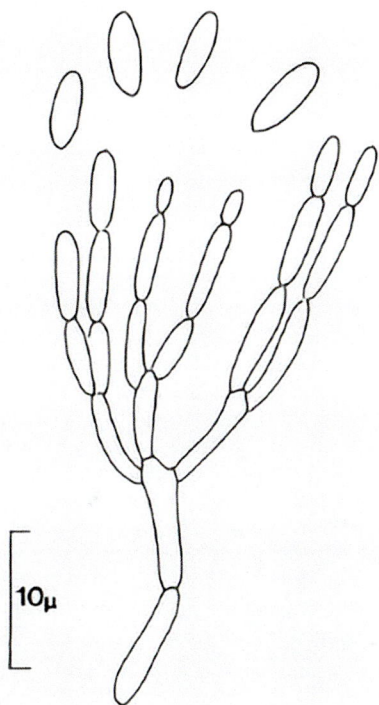

Fig. 3.23 *Metarhizium anisopliae* conidiophore and spores (CMI Descriptions of Pathogenic Fungi and Bacteria No. 609).

Fig. 3.24 Examples of *Trichoderma* colony morphology on 20% PDA (photo: K. Holmes).

Key references for identification:

Brady (1979b) CMI Descriptions of Pathogenic Fungi and Bacteria No. 609. *Metarhizium anisopliae*.

Trichoderma spp.

Colony morphology:

On PDA rapid growth is usually observed. Colonies are usually initially white with dark green conidia forming. Conidia formation can be dense, sparse or in concentric rings depending on the species. A yellow pigment is often observed diffusing into the media on the underside. A coconut odour can be produced by some species.
Microscopic view is shown in Fig. 3.25.

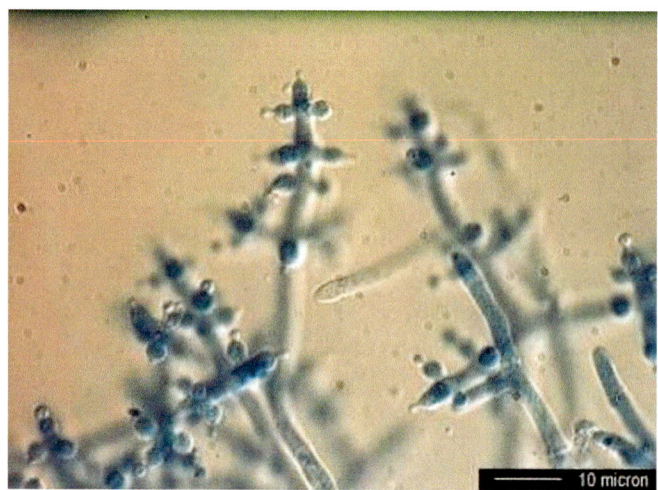

Fig. 3.25 Typical morphology of Trichoderma spp. (photo: K. Holmes).

Key references for identification:

Samuels (1996) *Trichoderma*: a review of biology and systematics of the genus. *Mycological Research*. 100:923–935.
Samuels, G.J. and Hebbar, P.K. (2015) *Trichoderma*: Identification and agricultural applications
Key contamination points are outlined in Table 3.1.

Table 3.1 Key control points for contamination checks.

Flow chart	Instructions	Important points/tips
10.10	• Ensure stock culture used for preparation of working cultures is free of contaminants and is the correct organism.	• Visual assessment of the culture.
10.50	• Check growing colony is the expected organism by checking growth characteristics of the colony and if required examine under the microscope.	
30.30	• Remove 0.5 ml of spore suspension on to 20% PDA or other media. • Use a glass spreader to spread over the surface of the agar plate. • Incubate for 3–5 days at 25°C. • Check for contamination.	• If contamination found, then products should be examined carefully.
30.50	• Visually check flasks for presence of contamination (cloudy, discoloration etc.). • Randomly select flasks and assess under the microscope. • Take sample and prepare plates as above in 30.30.	• If obvious contamination, then discard and do not use for inoculation. • If contamination found, then products should be examined carefully.
50.10	• Randomly select one autoclaved bag of solid substrate in a production run. • Prior to inoculation and just after autoclaving remove a few grains of the substrate and place on an agar plate, 20% PDA or other. • Incubate for 3–5 days at 25°C. • Check for contamination.	

3.4.6.2   Product quality control

A number of points can be addressed regarding the quality of the final product.

3.4.6.2.1   ESTIMATION OF VIABILITY OF SPORES

Instructions	Important points/tips
• Remove a small amount of conidia from the foil storage sachet and add to an open sterile Petri dish.	
• To rehydrate conidia, place Petri dish on platform over water in a sealed box for a minimum of 30 min.	
• Once rehydrated, take a small sample of conidia (on the tip of a clean micro-spatula) and make a dilute spore suspension (10^5–10 conidia/ml) (Samuels, 1996) in 9 ml Shellsol T.	
• Sonicate this suspension for 3 min to break up any chains of conidia.	• If no sonicator available, shake vigorously.
• Use a micro-spatula to transfer 2–3 drops of the conidia suspension onto the surface of a 50 mm Sabouraud dextrose agar (SDA) plate.	• SDA may be replaced by other agar media, PDA, PCA etc.
• The oil suspension should be spread carefully over the surface of the agar using the back of the micro-spatula and the plates kept at 25°C in an incubator for 24 h.	
• Remove from incubator after 24 h and count the proportion of germinated and non-germinated conidia by counting a minimum of 300 conidia per plate under the ×20 or ×40 objective of light microscope. Count all the conidia in each field of view.	
• Use a separate tally counter for germinated and non-germinated conidia. A total of 3 plates (300 conidia per plate) should be counted to give an accurate reading of the percentage germination.	
• To calculate the percentage germinated (percentage viability) carry out the following equation: *(total number of germinated conidia/total number of conidia)*100*	

3.4.6.2.2   DETERMINING NUMBER OF COLONY-FORMING UNITS OF A PRODUCT

Instructions
• Take a sample of dry conidia powder (approximately 0.1 g) using a small spatula and suspend the powder in 10 ml 5% sterile Tween 80 in a universal bottle. Replace the lid and shake vigorously or mix on whirlimixer.

Instructions

- Carry out a tenfold dilution series; transfer 1 ml of the spore suspension into a glass universal bottle containing 9 ml sterile water. Label this bottle −1 as this is the first one-in-ten dilution of the dilution series.
- Shake the −1 bottle vigorously, and then transfer 1 ml of this spore suspension to another universal bottle containing 9 ml sterile water. Label this bottle −2.
- Shake the −2 bottle and continue the dilution series down to −8.
- Now take a 0.2 ml sample from the −8 dilution and spread it over the surface of the agar in a Petri dish (90 mm diameter). Prepare two dishes for each of the dilutions in the series and label them accordingly.
- Incubate the dishes at 25°C in an incubator for approximately 3 days. Then remove the dishes and count the number of fungal colonies present across the dilution series. The average number of colonies per dish is calculated by adding the two counts together and dividing by two.
- Count the colonies as above for each pair of dilution dishes where individual colonies are still distinguishable. At the higher concentrations, the colonies will have merged together and it will not be possible to count the number of fungal colonies.
- To calculate the number of fungal conidia present in the original concentrated spore suspension, take the average counts from each of the dilutions where counts were possible. There will be a pattern in these values, in that the highest dilution should contain approximately one-tenth of the number of conidia in the dilution before and approximately 100th of the number of conidia than the dilution before this; for example, the average count of fungal colonies from the −8 dilution might have been 6, the count for the −7 dilution would therefore be expected to be about 60. In reality the count could be anything between 45 and 75. The −6 dilution would therefore contain between 450 and 750 colonies and it is quite likely that it would not have been possible to count these plates. The most accurate way of calculating the concentration of the original spore suspension in this example would be to take the average count from the −8 dilution and multiply it by 10, then add this number to the average count for the −7 dilution and divide by two. This will give an estimated average of the number of conidia in the −7 dilution. For example:
  - average count for the −8 dilution = 6
  - average count for the −7 dilution = 53
  - 6×10=60 (equivalent to the −7 dilution) so
  - 60 + 53=133. 133/2 = 56.5
- The calculation above gives the estimated number of colonies in 0.2 ml of the −7 dilution. To calculate the number of conidia in 1 ml of the −7 suspension, this must be multiplied by 5:
  - e.g. 56.5×5=282.5 colonies/ml
- To calculate the number of colonies in the concentrated solution, the colonies/ml must be multiplied by the dilution:
  - e.g. $282.5 \times 10^7 = 2.82 \times 10^9$ colonies(spores)/ml
- Divide this figure by the original weight to give the viable spores per gram (so $2.82 \times 10^9 / 0.1000 = 2.82 \times 10^{10}$/g).

The above method can also be used to assess the purity of the product. The number of non-product colonies (yeast, bacteria, and other fungi) as a percentage of the total can be calculated to provide a figure to indicate the purity of the final product.

3.4.6.2.3 ESTIMATION OF VIRULENCE OF PRODUCT (ENTOMOPATHOGENS). A bioassay can be carried out to ensure the product's efficacy (Fig. 3.27).

Use an alternative host such as Galleria melonella (wax moth) or field-collected insects.

Fig. 3.26 Rice grains inoculated with Trichoderma, left, and uninoculated, right (photo: S.Thomas, CABI UK).

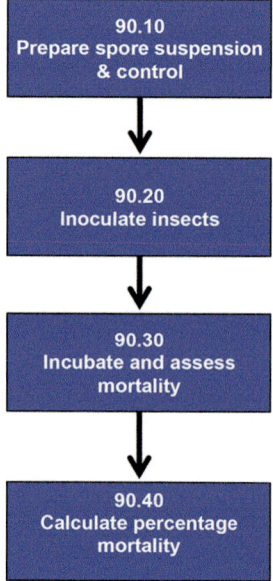

Fig. 3.27 Flow diagram for bioassay for entomopathogenic fungi.

Flow chart	Instructions	Important points/tips
90.10	● Prepare spore inoculum in vegetable oil (4×10^7 spores/ml).	● Use oil as a control. ● If sufficient material, could compare with spores from stock culture.
90.20	● Spray or apply drops of spore inoculum directly to the insects; in a sterile container apply inoculum to one set of 10 insects, with three replicates. ● Repeat for oil control.	
90.30	● Incubate at room temperature and assess mortality over 14 days.	
90.40	● Calculate percentage mortality after the treatment as follows: No. dead from spore treatment/no. of insects used × 100 Compare with control.	

3.4.7 Non-routine procedures

3.4.7.1 Maintenance of virulence: re-isolation from host (insect or pathogen)

For some fungal isolates constant subculturing can reduce the virulence of the isolate. To reduce the occurrence of this problem it can be useful to regularly passage the fungal isolate through a host. For the entomopathogenic fungi (*Beauveria* and *Metarhizium* spp.) an alternative host such as *Galleria melonella* (wax moth) can be raised in the laboratory and used, or field collections of target pests can be used instead.

For *Trichoderma* this is not a general problem however, re-isolation from a pathogen can be carried out.

3.4.7.1.1 MAINTENANCE OF VIRULENCE: ENTOMOPATHENIC FUNGI. The method used is a modification of the bioassay (Fig. 3.28).

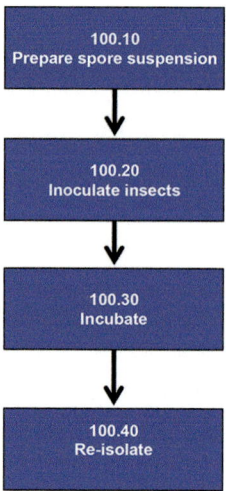

Fig. 3.28 Flow diagram for maintenance of entomopathogenic fungi virulence.

Flow chart	Instructions	Important points/tips
100.10	● Prepare spore inoculum in vegetable oil (4 × 10^7 spores/ml).	
100.20	● Spray or apply drops of spore inoculum directly to the insects; use a sterile container for one set of 10 insects, carrying out three replicates.	
100.30	● Incubate at room temperature, 25°C.	
100.40	● Select a colonized, sporulating insect cadaver and aseptically transfer spores to agar containing antibiotics, e.g. 20% PCA or PDA with Pen/Strep (see Annex 3.4.8.1.3). ● Incubate at 25°C and then carry out further subculture on to agar without antibiotics, to ensure the culture is pure.	● Once a pure culture is obtained prepare stock cultures for storage.

3.4.7.1.2 MAINTENANCE OF VIRULENCE FOR *TRICHODERMA* SPP.. Mode of action of *Trichoderma* spp. may vary. In general loss of virulence would not be expected; however, mycoparasitic *Trichoderma* spp. can be passaged.

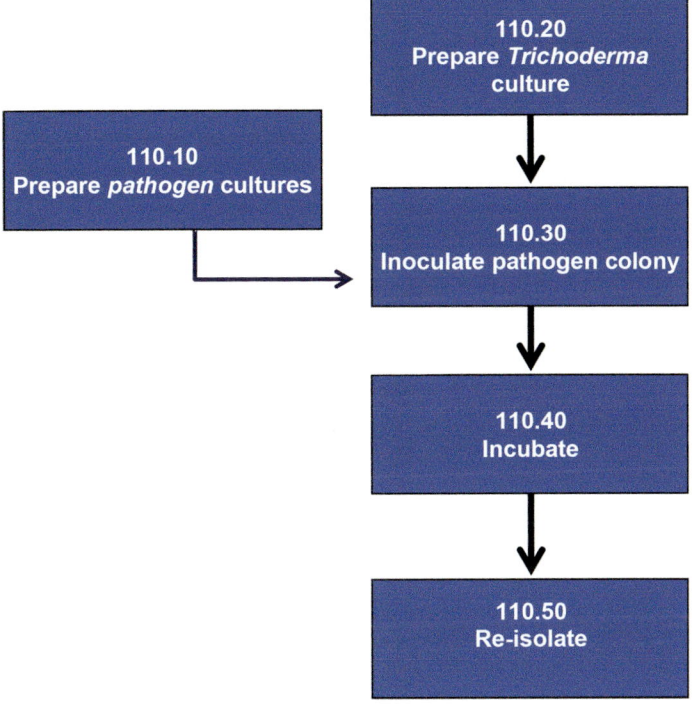

Fig. 3.29 Flow diagram for virulence maintenance of *Trichoderma*.

Flow chart	Instructions	Important points/tips
110.10	• Prepare pre-colonized plates of a known pathogen host/target of *Trichoderma*, on 9 cm agar plates (e.g. PDA). Prepare three replicates.	• Ensure plates are fully colonized by pathogen mycelium before use.
110.20	• Prepare *Trichoderma* cultures on agar plates.	
110.30	• Transfer 5 mm plugs from the actively growing leading edge of the *Trichoderma* culture and place at the edge of the pathogen colony (see Fig. 3.30).	• Ensure *Trichoderma* is only in contact with mycelium of the pathogen and not the underlying media.
110.40	• Incubate at 25°C for 10 days.	
110.50	• Remove agar plugs from the pathogen mycelium aseptically to 20% PDA with rose Bengal. • Where *Trichoderma* emerges from the plugs take a tip culture on to 20% PDA.	• Remove plugs from the opposite side of the original inoculum, in a direct line towards this original *Trichoderma* inoculum.

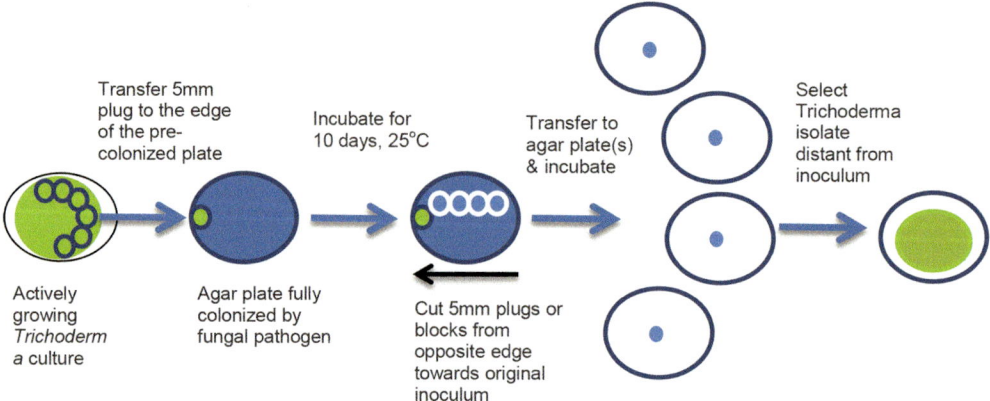

Fig. 3.30 Maintenance of virulence for *Trichoderma* spp.

3.4.8   Annexes

3.4.8.1   *Media and recipes (adapted from Ritchie, 2002)*

Media may be purchased as complete products or can be prepared from basic ingredients, as required. Agar is often available locally as undefined extract from seaweed.

3.4.8.1.1   MEDIA FOR FUNGAL SUBCULTURING/QUALITY CONTROL

Solid media

Sabaroud dextrose agar (SDA) - Beauveria/Metarhizium
Commercial product, as per manufacturer's instructions.

Local materials

Dextrose, 40 g
Peptone, 10 g
Agar, 20 g
Distilled water, 1000 ml

Method

- Add ingredients to water and dissolve in boiling water in beaker or saucepan.
- pH should be 5.6; adjust with NaOH if required.
- Autoclave at 121°C for 20 min.

Potato carrot agar (PCA) - Beauveria/Metarhizium/Trichoderma

Local materials

Potato, 20 g
Carrot, 20 g
Agar, 20 g

Tap water, 1000 ml

Method

- Wash, peel and grate vegetables.
- Boil for 1 h in 500 ml tap water.
- Strain through fine sieve retaining the liquid.
- Dissolve agar in 500 ml of tap water in a glass beaker or saucepan.
- Add strained liquid and mix.
- Autoclave at 121°C for 20 min.

Potato dextrose agar (PDA) - Beauveria/Metarhizium/Trichoderma
Commercial product, as per manufacturer's instructions

Local materials

Potatoes, 200 g
Dextrose, 15 g
Agar, 20 g
Tap water, 1000 ml

Method

- Scrub potatoes clean, but do not peel.
- Cut into 12 mm cubes.
- Weigh out 200 g.
- Rinse rapidly in running water.
- Place in 1000 ml of tap water and boil until soft (approximately 1 h).
- Put them through a blender.
- Add agar and dissolve in a glass beaker or saucepan.
- Add dextrose and stir until dissolved.
- Make up to 1000 ml.
- Agitate stock while dispensing, to ensure that each bottle has a proportion of solid matter.
- Autoclave at 121°C for 20 min.

If a weak vegetable decoction medium such as PCA is not available, commercial potato dextrose agar can be modified so that the sugar content is reduced to a minimum and can therefore be used as a substitute. For example, to make 25% strength PDA use 25% of the quantity of PDA powder recommended and add an extra 75% of the quantity of the equivalent brand of plain agar used for making water agar, i.e.:

25% strength PDA:

Potato dextrose agar powder, 9 g
Agar, 11.25 g
Distilled water, 1000 ml

20% strength PDA:

Potato dextrose agar powder, 7.2 g
Agar, 12 g

Distilled water, 1000 ml

20% PDA plus rose Bengal:

Potato dextrose agar powder, 7.2 g
Agar, 12 g
Distilled water, 1000 ml
Rose Bengal, 50 mg

20% PDA plus antibiotics:

Potato dextrose agar powder, 7.2 g
Agar, 12 g
Distilled water, 1000 ml
Penicillin G (1600 u/mg), 30 mg
Streptomycin sulphate (720 u/mg), 70 mg

3.4.8.1.2   MEDIA FOR LIQUID CULTURE

Potato dextrose broth (PDB) - Beauveria/Metarhizium/Trichoderma

Commercial product, as per manufacturer's instructions

From local materials

Potatoes, 200 g
Dextrose, 15 g
Tap water, 1000 ml

Method

Prepared as for PDA, without addition of agar.

Basal liquid medium (Krauss et al., 2002) - Trichoderma

Molasses, 80 g
Neopeptone, 10 g
Yeast extract, 2 g
Distilled water, 1 l

Method

Prepare as one batch in 2 l glass beaker or prepare 500 ml aliquots in 1 l flasks and autoclave. Autoclaving will ensure all materials are mixed. Autoclave 30 min at 121°C.

Molasses yeast extract medium (Papavizas et al., 1984) - Trichoderma

Molasses, 30 g
Brewer's yeast, 5 g
Distilled water, 1 l

Method

Autoclave 30 min at 121°C.

Brewer's yeast/sucrose broth – Beauveria/Metarhizium

Dried yeast, 20 g
Sucrose, 20 g
Tap water, 500 ml

Method

Autoclave 20–30 min at 121°C.

3.4.8.1.3   ADDITIVES AND ADDITIONAL LABORATORY MATERIALS

Antibiotics

Antibiotics should always be kept in a refrigerator. Stock solutions are prepared for use in agar media and may be sterilized by passing through a
bacterial filter (e.g. 20μ); although this is not usually necessary if aseptic
techniques are employed during preparation of stock solutions from an original sterile commercial preparation. Do not prepare large stock solutions as
potency can degrade quite quickly with some antibiotics.

Antibiotics should be added to 'hand-hot' media as excessive heat de-
natures most antibiotics.

Substances marked with an asterisk (*) are harmful, irritants, toxic or
carcinogens. ALWAYS read the label for safety precautions before use.

Penicillin G*

Use at 50–100 u/ml. Add to media after autoclaving. Active against
Gram-positive bacteria; should be used in conjunction with streptomycin
sulphate.

Streptomycin sulphate*

Use at about 50–100 u/ml. Add to media after autoclaving. Broad-spectrum
antibacterial but may also inhibit *Pythium* and some *Phytophthora* species.
Should be used in conjunction with penicillin.

Stain(s)

Rose Bengal
A stain that is effective against many bacteria and reduces the rate of spread
of fast-growing fungi. Used at about 50 mg/l and can be added before
autoclaving.

Sterilizing agents

Sodium hypochlorite solution
Suitable for surface sterilization.
Sodium hypochlorite* (full strength commercial bleach) (NaOCl), 10 ml
Distilled water, 90 ml
Detergent or Tween 80, few drops
Keep refrigerated to prevent rapid loss of active chlorine. Usable while it
smells fairly strongly of chlorine.

3.4.8.2   Maintenance of stock cultures (adapted from Smith, 2002)

The number of subcultures of a working culture should ideally be minimized to reduce the potential for contamination and to ensure that virulence of the product is maintained. It is therefore good practice to maintain stock cultures, in addition to working cultures. A range of methods can be utilized to maintain fungal cultures, depending on facilities available (Fig. 3.31).

Dry storage

- Agar (PDA etc.) slopes are prepared in glass Universal (or McCartney) bottles.
- Agar slopes are inoculated with plugs or tissues from the leading edge of a growing colony.
- The cultures are incubated at 25°C until the slope is fully colonized by the fungus.
- Mature healthy cultures are stored with their caps loose in racks at 15–18°C.
- Duplicates may also be stored in a refrigerator at 4–7°C.

Water storage

- 10 Agar blocks (6 mm^3) or plugs (5 mm diameter) are cut from the growing edge of a fungal colony.
- The blocks are placed in sterile distilled water in 5 ml glass bijou bottles and the lids are tightly screwed down; they are stored at 20–25°C.

Fig. 3.31 Storage in water and on dry agar slope (photo: K. Holmes).

- 10 replicates are prepared.
- Retrieval is by removal of a block and placing, mycelium down, on a suitable growth medium.
- Duplicate batches can be maintained in a refrigerator at 4–7°C.
- Storage periods of 2–3 years have been obtained.

Oil storage

Covering cultures on agar slants (30° to the horizontal) in 30 ml universal glass bottles with mineral oil prevents dehydration and slows down the metabolic activity and growth through reduced oxygen tension.

- Agar (PDA etc.) slopes are prepared in glass Universal (or McCartney) bottles.
- Agar slopes are inoculated with plugs or tissues from the leading edge of a growing colony.
- The cultures are incubated at 25°C until the slope is fully colonized by the fungus.
- Mature healthy cultures are covered by 10 ml of sterile mineral oil (liquid paraffin or medicinal paraffin specific gravity 0.830–0.890 previously sterilized by autoclaving twice at 121°C for 15 min).
- Universal bottles are stored with their caps loose in racks at 15–18°C.
- Duplicates may also be stored in a refrigerator at 4–7°C.
- Retrieval from oil is by removal of a small amount of the colony on a mounted needle, draining away as much oil as possible and then streaking on to a suitable agar medium.
- Growth rate can often remain restricted due to adhering oil, so subculturing must be done by re-isolating from the edge of the colony and transferring to fresh medium.
- Inoculating an agar slope centrally sometimes gives better results as excess oil can drain down the slope, allowing the fungus to grow more typically towards the top.

Although some fungi can survive for up to 40 years using this method, some may deteriorate and require subculturing every 2 years.

Silica gel storage

Sporulating fungi have been stored for 7–18 years in silica gel with good revival and stability.

- Fill universal glass bottles one-third full with medium grain plain 6–22 mesh non-indicating silica gel and sterilize with dry heat (180°C for 3 h).
- Place bottles in a tray of water and place in a deep freeze (−20°C).
- Prepare spore suspensions in cooled 5% (w/v) skimmed milk.
- Add the suspension to the cool gel to three-quarters wet it (approximately 1 ml) and agitate to distribute the spore suspension throughout.

- Store the bottles with the caps loose for 10–14 days at 25°C until the silica gel crystals dry and separate easily.
- Screw the caps down and store the bottles at 4°C (though storage between 20 and 25°C is satisfactory) in air-tight containers over indicator silica gel to absorb moisture.

Retrieve the strains by scattering a few crystals on to a suitable medium.

Other more effective long-term storage methods can be used, including lyophilization and cryopreservation. These, however, require more technical experience and specialist equipment and so are not discussed further here.

3.4.8.3   Haemocytometer use

See Fig. 3.32–Fig. 3.35.

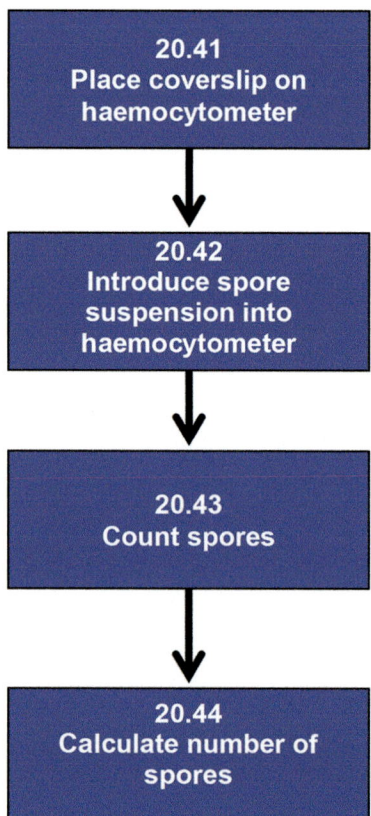

Fig. 3.32 Flow diagram for counting spores using a haemocytometer.

Flow chart	Instructions	Important points/tips
20.41	• Place the glass cover slip over the counting chamber, slowly force the coverslip down and move slowly back and forth over the chamber until a seal is formed.	• The cover slip is in position when Newton's rings can be observed where the coverslip meets the haemocytometer surface. This ensures the volume of the counting chamber is correct. • Be careful not to apply too much pressure as the coverslip will break.
20.42	• Using a Pasteur or automatic pipette, introduce a spore sample into the counting chamber. Place the edge of the pipette tip at the edge of the coverslip on the central chamber. • Slowly release the spore sample, which will be taken up into the counting chamber. Stop once filled.	• Do not overfill as this may remove the coverslip. If bubbles appear or the coverslip moves start again from 20.41.
20.43	• Place the loaded haemocytometer on a compound microscope platform. Focus the microscope (× 400) so the cells can be seen on the counting grid. • Count the cells in five big squares (see Fig. 3.33 below) for one chamber then repeat for second chamber.	• A general rule when counting cells is that those touching the upper and left sides of the grid should be counted, while those touching the right and lower edges should be excluded from the count (see Fig. 3.34)

Flow chart	Instructions	Important points/tips
20.44	• **Haemocytometer calculation [for improved Neubauer 1/16, 1/400 mm^2]** • To calculate the number of spores in 1 cm of original solution: • For a small square = X × 4 × 10^6 • For a medium size square = X × 2.5 × 10^5 • For a large square = X × 1 × 10^4 • Count number of spores in five medium size squares, take the mean. Adjust concentration in original suspension by dilution factor (if any used). A more detailed calculation example:	• This will provide the number of spores. However, if a specified weight added, divide this figure by the original weight of spores added to the stock suspension to give the number of spores per gram.

	Grid 1	Grid 2
	66	68
	67	48
Spores counts in grids	67	84
	51	64
	75	55
TOTAL	**326**	**319**
Average	**322.5**	
Spores per ml of dilution (−2)	=322.5 × 5 × 10^4	=1612.5 × 10^4
Spores per ml stock	=1612.5 × 10^4 × 10^2	=1612.5 × 10^6
		=1.6125 × 10^9
Spores per ml in 10 ml stock	=1.6125 × 10^9 × 10^1	=1.6125 × 10^{10}
Spores per gram	=1.6125 × 10^{10}/0.1018 (original spore weight measured into stock)	**=1.43 × 10^{11} spores per g**

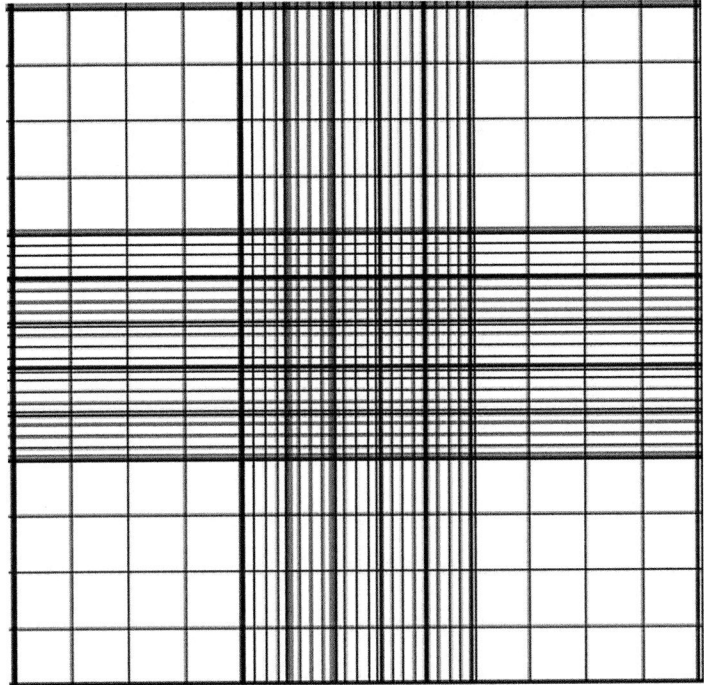

Fig. 3.33 Counting grid of haemocytometer showing large, medium and small counting squares.

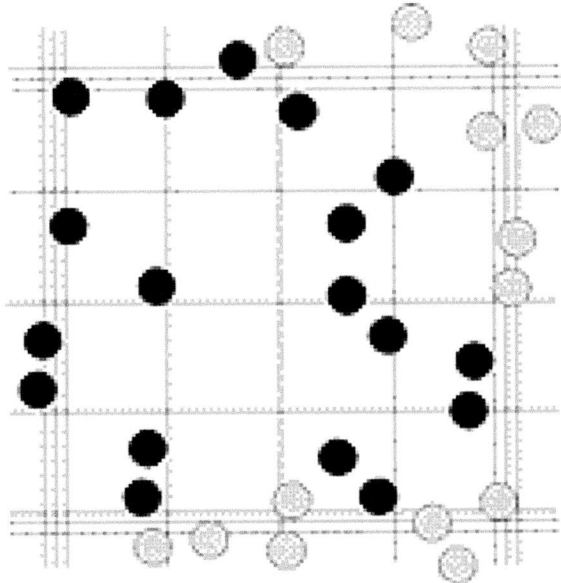

Fig. 3.34 Cells to count in large Neubauer chamber (black cells).

3.4.8.4   Equipment required for the mass production of fungal biocontrol agents

Equipment

- Petri dishes (glass or plastic), 9 cm or 5 cm
- Autoclave
- Rotary shaker
- Incubator
- Refrigerator
- Balance (to three decimal places small mass and large 1 kg mass)
- Bunsen or alcohol burner
- Compound microscope
- Haemocytometer
- Glassware (500 ml beakers,1000 ml, 500 ml, 250 ml conical flasks and/ or medical flats, 500 ml, 100 ml, 10 ml measuring cylinders, 25 ml glass or plastic universal containers, funnels)
- Hand-held counters
- 1 ml, 10 ml glass Pasteur pipettes or automatic (e.g. Gilsen) pipettes
- Spatulas
- Plastic trays
- Thermometers (incubator and room)
- Water distiller
- Metal sieve (300 μm mesh)
- Desiccator
- Small oven

Consumables

- Potato dextrose agar
- Technical agar
- Molasses
- Brewer's yeast
- V8 broth
- Potato dextrose broth
- Mineral oil
- Bleach (sodium hypochlorite)
- Aluminium foil
- Non-absorbent cotton wool
- String
- Elastic bands
- Ethanol
- Syringes and needles
- Weighing boats
- Autoclave tape
- Parafilm
- Distilled water

- Autoclavable plastic bags
- Marker pens (waterproof)
- Silica gel (non-indicating)
- Tween 80 (or other surfactant/soap)

3.4.8.5 Mass production timeline

The mass production timeline presented here (see Fig. 3.36) assumes the initial starting point is the use of a stock culture. If initiating from an existing working culture, which ideally should not be subcultured more than five times for the entomopathogens, then the initial phase may be considerably shorter. Similarly, this production includes the production of a liquid starter culture. In some cases, depending on the volumes to be produced, it may be possible to use spore inoculum directly from agar plate cultures. This would also reduce the production time.

3.4.8.6 Indicative production of spores

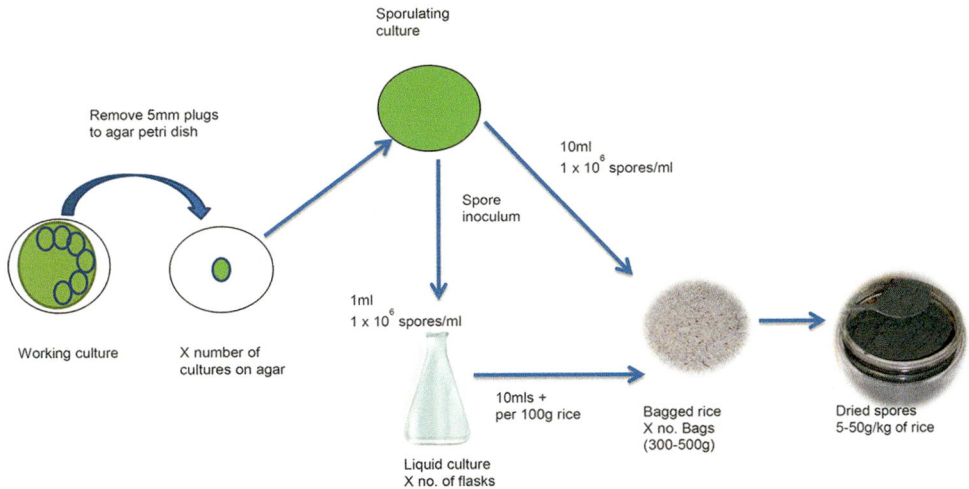

Fig. 3.35 Production flow from working culture to product (spores).

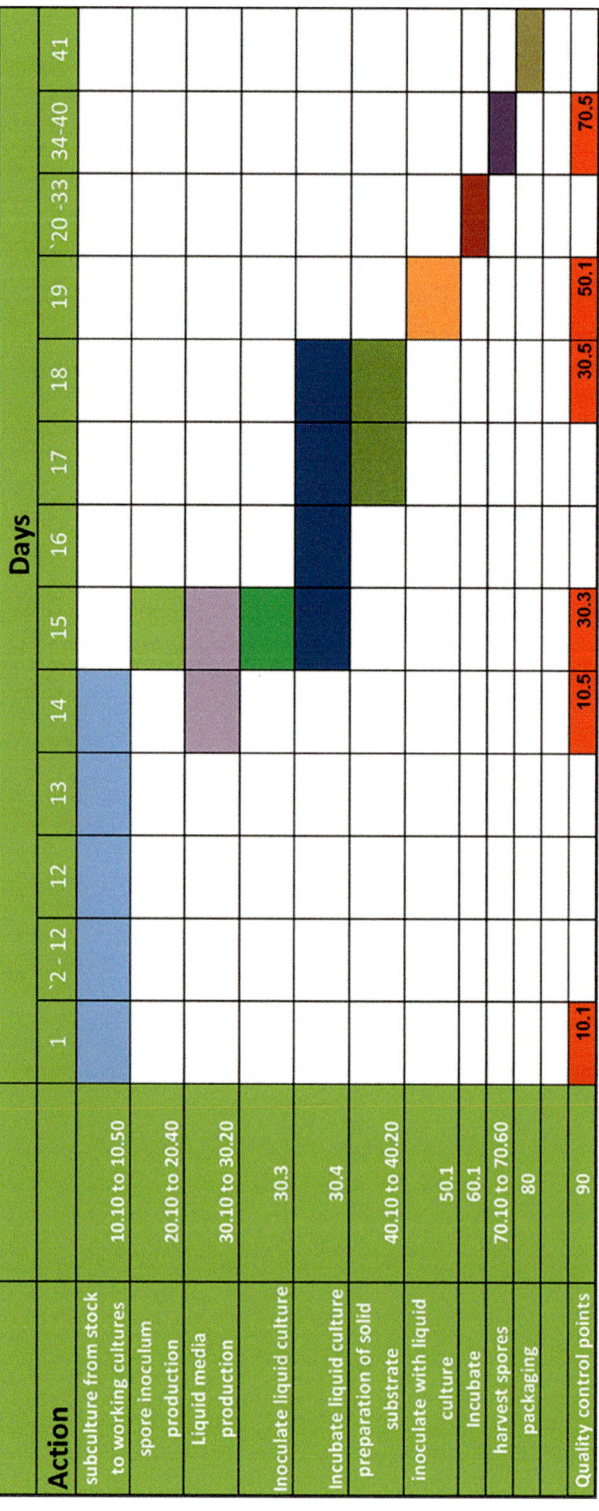

Fig. 3.36 Mass production timeline.

References

Brady, B.L.K. (1979a) CMI Descriptions of Pathogenic Fungi and Bacteria No. 602 In: *Beauveria bassiana*. Commonwealth Mycological Institute, Kew, UK.

Brady, B.L.K. (1979b) CMI Descriptions of Pathogenic Fungi and Bacteria No. 609 In: *Metarhizium anisopliae*. Commonwealth Mycological Institute, Kew, UK.

Grossrieder, M., Zheng, L., Song, K., Pyon, Y.C. and Jo, R. (2009) *Trichogramma Rearing Manual, Biological Control of the Asian Corn Borer in DPR Korea*. CABI International, Europe – Switzerland.

Krauss, U., Martínez, A., Hidalgo, E., ten Hoopen, M. and Arroyo, C. (2002) Two-step liquid/ solid state scaled-up production of *Clonostachys rosea*. *Mycological Research* 106(12), 1449–1454.

Lokanadhan, S., Muthukrishnan, P. and Jeyaraman, S. (2012) Neem products and their agricultural application. *Journal of Biopesticides* 5(Supplementary), 72–76.

Lomer, C.H. and Lomer, C.J. (1998) Mass production of fungal pathogens for insect control In: *Insect Pathology Manual*. CABI Bioscience, pp. 209–228.

Papavizas, G.C., Dunn, M.T., Lewis, J.A. and Beagle-Ristaino, J. (1984) Liquid fermentation technology for experimental production of biocontrol fungi. *Phytopathology* 74(10), 1171–1175.

Ritchie, B.J. (2002) Mycological media and methods (2002) In: J. M. Waller, J. M. Lenné, and S. J. Waller (eds) *Plant Pathologists' Pocketbook*. CABI Publishing, Wallingford, UK, pp. 410–431.

Samuels, G.J. (1996) *Trichoderma*: a review of biology and systematics of the genus. *Mycological Research* 100(8), 923–935.

Samuels, G.J. and Hebbar, P.K. (2015) *Trichoderma: Identification and agricultural applications (No. LC-0862)*. The American Phytopathological Society.

Smith, D. (2002) Culturing, preservation and maintenance of fungi In: J. M. Waller, J. M. Lenné, and S. J. Waller (eds) *Plant Pathologists' Pocketbook*. CABI Publishing, Wallingford, UK, pp. 384–409.

Vijayalakshmi, K., Radha, K.S. and Shiva, V. (1995) *Neem: A user's manual*. Centre for Indian Knowledge systems, Chennai.

4 Training Guide for Field Technicians and Farmers: Biopesticides and How to Work with Them

4.1 Introduction to Biopesticides

Biopesticides, also referred to as biocontrol agents, are an integral part of many IPM/ICM strategies for the control of pests. They are often deployed to control insect pests but may also be used to target microbial pathogens, nematodes or weeds. Biocontrol agents are a promising avenue for development in that they present several advantages over conventional chemical pesticides.

Main groups of biopesticides (including macrobials):

- **Macrobials** (macroorganisms) include insect natural enemies (e.g. *Trichogramma*) and entomopathogenic nematodes (although these are often considered under microbials).
- **Microbial biopesticides** (microorganisms) may be either microorganisms (e.g. bacteria, fungi, viruses, viroids or protozoa) or their products (metabolites, e.g. protein toxins) as the active substance. Entomopathogenic nematodes are often classed as microbial pesticides.
- **Biochemical biopesticides** are a diverse group that could be considered to include naturally derived biochemicals such as **plant extracts/botanicals**, which are materials derived from plants that are active against the target pest or pathogen. These may have direct effects on the target pest or indirect effects via the host plant. **Semiochemicals** are naturally occurring chemicals emitted by plants, animals and other organisms that modify insect pest behaviour. They may also be produced synthetically. Semiochemicals can be used as repellants, attractants for traps, or for mating disruption. Biochemical pesticides can also include metabolites derived from fermentation of living microorganisms, e.g. Spinosad.

Often farmers and other technical staff may not appreciate that biocontrol agents can be effective replacements for CPAs. This training topic looks at providing a brief overview of the advantages of biocontrol agents and how they differ from conventional chemical CPAs.

Topics to be covered in this section:

- What are biopesticides?
- What are the advantages of biopesticides over CPAs?
- How do biopesticides differ from conventional CPAs?

4.1.1 ACTIVITY: What are biopesticides?

Time required: 45 min
Target audience: Farmers and field technicians
Location: Classroom

MATERIALS:

- Flipchart and pens
- Handout of types of biopesticides

PROCEDURES:

Use a flipchart or other visual aids to lead a discussion with trainees.

- What do the trainees understand by the term biopesticide?
- Can they provide a list of the main groups of biopesticides?
- Can they give examples of each of these groups and their target pest?
- Have they experience of using biopesticides? If yes what was their impression?
- What are the advantages of biopesticides over CPAs?
- What are the potential disadvantages of biopesticides compared with CPAs?

Some possible examples of advantages and disadvantages are listed in the table below.

Advantages of biopesticides	Disadvantages of biopesticides
Usually inherently less toxic than conventional pesticides	Slow speed of action (not always)
Lower potential for residues	High specificity (not always)
Can be cheaper than chemical pesticides (if locally produced)	Can be more expensive commercial products
Often effective in small quantities and decompose quickly	Variable efficacy due to biotic and abiotic factors
May be more effective in the long term	Require careful storage and handling
Less prone to development of resistance in target	
Generally more specific, only affecting target organism and closely related organisms	
Potential for long-term self-dissemination/ maintenance (classical biocontrol)	

Advantages of biopesticides	Disadvantages of biopesticides
Lower non-target effects	
Use in IPM system can maintain yield while reducing use of conventional pesticides	
Often less onerous registration requirements	

4.1.2   ACTIVITY: Pest and biopesticide matching

Time required: 45 min
Target audience: Farmers and field technicians
Location: Classroom

MATERIALS:

- Flipchart and pens
- Pictures of major pests (e.g. hornworm, budworm) and diseases, or symptoms of pest damage (e.g. diseased plants or leaves with flea beetle damage)
- Biopesticide product labels and/or pictures of biopesticide organisms

PROCEDURES:

Have enough pictures and containers so that half of the group can have a picture of a pest and the other half of the group can have a product label (and/or a picture of the biopesticide). The pictures should be of pests that the farmers will be familiar with.

Distribute the pictures and product labels. Tell the participants to look for their 'match' (i.e. the participants holding pictures of biopesticides should find at least one person holding a picture of a pest that the biopesticide works against; participants holding pictures of pests should look for people holding biopesticides that are effective against that pest).

Once everyone has a partner, ask the pairs to form groups according to biopesticide type (i.e. all insect pests and entomopathogens should group together).

4.1.3   ACTIVITY: How do biopesticides differ from conventional pesticides?

Time required: 45 min
Target audience: Farmers and field technicians
Location: Classroom

MATERIALS:

- Flipchart and pens

- Biopesticide product labels
- Conventional CPA product labels

PROCEDURES:

Ask the participants to divide into groups. Give each group an example of a label for a biopesticide product and a conventional CPA product. Ask the groups to read through the labels and to make a note of ways in which they differ (or are the same) for each of the categories of information listed in the table below.

Category of information	Biopesticide product	Conventional CPA
Name		
Active ingredient		
Toxicity class (band colour)		
PPE		
REI		
PHI		
Preparation		
Environmental conditions for application		
Equipment		
Application timing		
Application interval		
Maximum number of applications		
Compatibility with other CPAs		
Storage requirements		

Once the participants have finished their review, ask each group to present their findings to the rest of the group in plenary. Discuss with the group the ways in which the biopesticides differ from the conventional chemical CPAs. Depending on the products selected, some key differences could include the toxicity class (on average, biopesticides will have lower toxicity profiles than chemical CPAs), the PPE requirements (often a lower level of protection is required for biopesticides application but some PPE is still needed), re-entry interval (REI and PHI typically lower), etc.

Wrap-up the activity by emphasizing the benefits of using biopesticides over conventional CPAs. These include;

- Reduced environmental impact
- Reduced acute and chronic effects on the health of those applying the biopesticide and the surrounding community
- Reduced potential for development of resistance
- Reduced impact on natural enemies

4.2   Managing Pathogens of Tobacco with Biopesticides

4.2.1   *Trichoderma* for the management of soil-borne pathogens in tobacco

BACKGROUND

- **Active substance:** *Trichoderma* spp.
- **Substance group:** Microbial – Fungus
- **Target pests:** Many significant soil-borne pathogens of tobacco including *Pythium* spp., *Phytophthora* spp., *Thielaviopsis basicola*, *Rhizoctonia solani*, *Fusarium* spp., *Verticillium* spp.
- **Application site and/or stage:** Seedbed
- **Special considerations for use:** If applying to external seedbeds, apply during late evening to avoid high temperatures during initial establishment. Can be incorporated into growth substrate or applied as seed treatment. As with all biological agents wear appropriate PPE, as indicated on product label.
- **Overview of the regulatory status:** *Trichoderma* spp. are registered worldwide. Check the latest list of nationally registered pesticides to make sure that it is allowed for use in tobacco.

Trichoderma are fungi that are known to be present in most soils globally. They are also found within host plant tissues as endophytes (beneficial symbionts).

Trichoderma species have been extensively researched as biological control agents of fungal pathogens, and have been shown to control most fungal pathogens studied, although the particular species or strain and its efficacy may vary for any given fungal pathogen.

The mechanisms employed by *Trichoderma* species include:

- Mycoparasitism
- Antibiosis
- Competition for nutrients or space
- Induced resistance
- Improvement of nutrient utilization. Overall increased plant health may improve crop vigour, yields and quality, especially under stressful conditions.

Trichoderma spp. have been studied extensively as biocontrol agents in soil-based systems and for aerial application, and are known to as act as opportunistic plant symbionts. *Trichoderma* may demonstrate rhizosphere competence; the ability to colonize the area, usually less than 1 cm, around the plant roots. It has also been demonstrated that *Trichoderma* species colonize the root tissues, where they can induce local or systemic resistance within the host plant. *Trichoderma* species are known to enhance plant health by improving root growth, removing potential plant pathogens from the rhizosphere, and also enabling nutrient uptake by plants.

In addition, *Trichoderma*'s innate resistance to agricultural chemicals such as copper and sulphur means it may be used in conjunction with these.

The use of *Trichoderma* in nurseries and glasshouse systems is well established and it provides important protection to the early stages of plant production. For example, in plant nurseries it can provide protection to the seed/seedling from damping-off pathogens to enhance production. In addition, those strains of *Trichoderma* with rhizosphere competence or endophytic ability can in principle remain with the host plant when transplanted to the field and provide further direct protection of the root zone, increased stress tolerance (e.g. drought) and aboveground protection against pathogens through induced resistance.

Trichoderma can be incorporated into growth substrates directly as a solid substrate/powder/granules or as a solution diluted in water or oil (depending on the hydrophobicity of the spores). It can be applied using a knapsack sprayer to the substrate surface. Application rates can vary (see product label for application rate).

Trichoderma products should be stored in a cool, dry location and used within the recommended time-frame (check product label for details of storage conditions and expiry date).

Topics covered in this section include:

● Recognizing *Trichoderma*
● Demonstration of *Trichoderma* disease control
● Demonstration of *Trichoderma* enhancement of plant growth

4.2.1.1 *ACTIVITY: Understanding what* Trichoderma *is*

Time required: Initially 30 min. Subsequent observation over 5 days (or pre-prepare as visual demonstration)
Target audience: Farmers and field technicians
Location: Classroom

MATERIALS:

● Commercial or local *Trichoderma* product (spores)
● Potato or apple
● Knives
● 70% alcohol
● Clean containers
● Non-absorbent cotton wool
● Printed table
● Flipchart and pens

PROCEDURES:

Farmers and other technical staff may not appreciate that biocontrol agents can be effective replacements for CPAs. This training topic demonstrates *Trichoderma* as a suitable biocontrol agent for the control of fungal diseases (though it may have indirect effects on other diseases). This activity will provide trainees with an understanding of what *Trichoderma* is and the benefits it can provide.

Prepare a flipchart or poster illustrating the key points about *Trichoderma* as above. Discuss with trainees their knowledge and experience of *Trichoderma*.

Day one:

- Clean two containers with alcohol
- Cut a potato into two halves
- Place a small quantity of the *Trichoderma* spores at one edge of the surface of one of the potato halves.
- Place in a sealed, but not air-tight, container
- Place the other uninoculated potato half in the second clean container
- Keep both in a cool dark place

Day three:

- Observe the cut potato surfaces of the *Trichoderma*-inoculated and uninoculated control
- Write observations in the table supplied

Day five:

- Repeat the above steps for day three.

Treatments	Colour			Texture		
	0d	3d	5d	0d	3d	5d
Potato with spores						
Potato without spores						

QUESTION FOR DISCUSSION

- What was observed and what could this mean for control of diseases?

Note. Facilitators: Alternatively, set up demonstration materials by preparing staggered inoculations with Trichoderma, with controls. Then present all the growth stages to the trainees and request their observations.

4.2.1.1.1   ADDITIONAL ACTIVITY.

PROCEDURE:

If a laboratory and materials are available the following could also be prepared to provide further information to the trainees.
Prepare slides of the *Trichoderma*.

- Prepare 5 cm or 9 cm agar plates containing 10% potato dextrose agar.

- Inoculate one edge of the plate with *Trichoderma* spores.
- Incubate at 25°C in the dark for 4–7 days.
- When the *Trichoderma* has colonized two-thirds of the plate and is sporulating, remove a small amount of mycelia and place into a drop of cotton-blue stain. Tease the mycelia apart and place a cover slip over the mycelia.
- View under the microscope to identify conidiophore and conidia.
- If the slide is useful for demonstration then seal the edge with nail varnish.
- Prepare a number of slides for demonstration.
- Have participants take turns looking at *Trichoderma* under the microscope and ask them what they see.
- Explain to them that what they are seeing is the living organism of *Trichoderma*.
- Prepare photo sheets illustrating conidiophores and spores of *Trichoderma*.

4.2.1.2   ACTIVITY: Assessment of control of damping-off with Trichoderma

Time required: 21 days
Target audience: Farmers and field technicians
Location: Classroom

MATERIALS:

- Commercial or local *Trichoderma* product (spores)
- Seedling trays or pots
- Compost or soil
- Mixing bowl
- Seed
- Plastic labels
- Marker pens
- Petri dish
- Filter paper/tissue paper
- Handout illustrating mechanisms and target pathogens
- Flipchart and pens

PROCEDURES:

Using a flipchart or poster discuss with farmers the main targets of *Trichoderma*. Discuss how they think *Trichoderma* may control diseases.

Discuss each of the mechanisms mentioned below and provide an example of each:

- Mycoparasitism, e.g.
- Antibiosis, e.g.
- Competition for nutrients or space, e.g.
- Induced resistance, e.g.

After the discussion, split the group into pairs for a practical session to prepare a *Trichoderma* treatment and a control treatment. Each pair will prepare their own materials.

Preparation of the Trichoderma treatment:

- Ask the pairs to read the product label to determine the number of spores per gram of product.
- Mix sufficient soil (or compost) for one tray or 20 pots with the *Trichoderma* product to provide a final concentration of 1×10^6 spores per gram of soil (or, if there is sufficient time and materials, prepare a range of concentrations, e.g. 1×10^5, 1×10^6, 1×10^7). Ensure the soil is not dry, but not too wet either.
- Place soil/*Trichoderma* in tray/pots (increase number of pots or replicates as required).
- Leave for 2 days (can be used directly if time is an issue).

Preparation of the control treatment:

- At the same time as the *Trichoderma* treatment is prepared, set up the same number of pots for an uninoculated control.
- After 2 days, sow one seed into each pot or compartment of the seed tray.
- Prepare a seed control, by placing 100 seeds in a petri dish on moist tissue paper or filter paper.
- Water as required.
- After 7–10 days count the number of emerged seedlings and calculate the percentage emergence (record which pots or compartments have seedlings).
- Assess the number of germinated seeds in each petri dish.
- Reassess at 14 and 21 days, recording the position of emerged and damped-off plants (post-emergence).
- Calculate pre- and post-emergence damping-off:

    - Pre-emergence damping-off (those seedlings failing to emerge)
    - Post-emergence damping-off (seedlings that emerge and subsequently die).

Treatments	No. of seeds sown	No. of plants emerged	% plant emergence	% pre-emergence damping-off	% post-emergence damping-off
Trichoderma treated					
Control					

Question: Is there a difference between the treated and untreated pots?

Note. Facilitator could pre-prepare all the above as a demonstration if time is limited.

Facilitator should carry out a pre-assessment to ensure that the *Trichoderma* product is effective and identify an appropriate concentration to use to provide a clear effect.

Note. This activity could be a combined activity with the one outlined below, to show the effect on growth at the same time.

4.2.1.3   ACTIVITY: Assessment of growth enhancement using Trichoderma

Time required: 30 days (or pre-prepare materials as a demonstration)
Target audience: Farmers and field technicians
Location: Classroom

MATERIALS:

- Commercial or local *Trichoderma* product (spores)
- Seedling trays or pots
- Compost or soil
- Mixing bowl
- Seed
- Plastic labels
- Marker pens/pencils
- Petri dish
- Filter paper/tissue paper
- Handout of mechanisms and target pathogens
- Flipchart and pens

PROCEDURES:

Using a flipchart or poster ask the trainees to list the potential additional benefits of using *Trichoderma*.

Discuss each of the mechanisms mentioned below and provide an illustration of each.

- Solubilization and sequestration of inorganic nutrients
- Induced tolerance to abiotic stress
- Increased growth

After the discussion, split the group into pairs for a practical session to prepare a *Trichoderma* treatment and a control treatment. Each pair will prepare their own materials.

Preparation of the Trichoderma treatment

- Read the product label to determine number of spores per gram of product.
- Mix sufficient soil (or compost) for one tray or 20 pots with the *Trichoderma* product to provide a final concentration of 1×10^6 spores per gram of soil (or, if there is sufficient time and materials, prepare a range of concentrations, e.g. 1×10^5, 1×10^6, 1×10^7). Ensure soil is not dry, but not too wet either.

- Place soil/*Trichoderma* in trays/pots (increase number of pots or replicates as required).
- Leave for 2 days (can use directly if time is an issue).

Preparation of the control
- Prepare the same number of pots for the uninoculated control, at the same time as the Trichoderma treatment.
- After 2 days (or immediately) sow one seed into each pot or compartment of seed tray.
- Prepare a seed control, by placing 100 seeds in a petri dish on moist tissue paper or filter paper.
- Water as required
- After 30 days carefully remove 20 plants.
- Remove soil by carefully washing under running tap or in a bowl/bucket.
- Assess the root length and shoot length for each plant and note the average in the table below. Do this for each treatment then calculate the percentage change for the Trichoderma treatment compared to the control.
- Remove the shoot by cutting at the collar and weigh the total root mass. Do this for each treatment then calculate the percentage change for the Trichoderma treatment in comparison to the control.

	Root length (cm)	Shoot length (cm)	Root mass (g)
Trichoderma treatment			
Control			
% change			

Question: Is there a difference between the treated and untreated plants? If yes, why do the trainees think this has occurred? What benefits would this provide for the plant?

Note. Facilitator could pre-prepare as a demonstration if time is limited.

The facilitator should carry out a pre-assessment to ensure that the *Trichoderma* product is effective and identify the appropriate concentration to use to demonstrate a clear effect.

4.2.2 *Ampelomyces quisqualis* for the management of powdery mildew

BACKGROUND

- **Active substance:** *Ampelomyces quisqualis*
- **Substance group:** Microbial – Fungus
- **Target pests:** Powdery mildew (*Erysiphe* spp.)
- **Application site and/or stage:** Foliar application

- **Special considerations for use:** Apply in the evening or early in the morning. Do not apply in full sun. As with all biological agents wear appropriate PPE, as indicated on product label.
- **Overview of the regulatory status:** *Ampelomyces quisqualis* is registered for use in tobacco in a few countries

Ampelomyces quisqualis is a fungus found widely in nature, which selectively hyperparasitizes powdery mildew (*Erysiphe* spp.), suppressing its development. It spreads through sporulation and is self-propagating.

Products containing *A. quisqualis* can work very effectively in a spray programme to prevent or to control powdery mildew. Control works best if *A. quisqualis* is applied when powdery mildew infestation is low (below 3%), so the weather should be monitored for the warm, humid conditions that are conducive to powdery mildew.

Ampelomyces quisqualis can be applied using standard spray application techniques. It requires high levels of humidity, so it is recommended that *A. quisqualis* is applied early in the morning or late in the afternoon when there is likely to be dew on the tobacco leaves. Mineral oil adjuvants can also be used to help to ensure high enough humidity levels for spore production and hyphal invasion.

Ampelomyces quisqualis should be stored in a cool, dry place, preferably under refrigeration. Refrigeration will significantly extend the shelf life of products containing *A. quisqualis*. Exposure to extreme temperatures will kill the *A. quisqualis* spores, rendering the product useless. Once the product has been opened, it should be used completely.

4.2.3 *Coniothyrium minitans* for management of *Sclerotinia* spp.

BACKGROUND

- **Active substance:** *Coniothyrium minitans*
- **Substance group:** Microbial – Fungus
- **Target pests:** *Sclerotinia* spp.
- **Application site and/or stage:** soil application (generally 2–3 months prior to crop)
- **Special considerations for use:** Apply in the evening or early in the morning. Do not apply in full sun. As with all biological agents wear appropriate PPE, as indicated on product label.
- **Overview of the regulatory status:** *Coniothyrium minitans* is registered for use in tobacco in some countries.

Coniothyrium minitans is a fungus that commonly occurs in soils worldwide. It is a specialized biological control agent that targets the fungal pathogen *Sclerotinia sclerotiorum* and other *Sclerotinia* spp. (causal agents of white mold). *Coniothyrium minitans* is a mycoparasite of the sclerotia (overwintering structures of the fungal pathogen) of *Sclerotinia* spp. *Sclerotinia* can survive overwinter without its host or host plant residues as sclerotia (hard mycelial bodies) in the soil. The *C. minitans* hyphae parasitize the sclerotia of the *Sclerotinia* spp., and then they produce spores that are released into the environment to continue the infection cycle.

Given that it targets the overwintering structures of *Sclerotinia*, *C. minitans* should be applied at the end of the crop cycle or prior to the next transplanting or sowing of tobacco. This will ensure that the sclerotia in the soil are destroyed before they can germinate and infect the next crop.

Use higher application rates when the weather conditions are expected to be conducive for disease development, or if the field has a history of disease and the disease pressure is high. *Coniothyrium minitans* has no curative effect and therefore is not effective against plants already infected with disease at the time of application.

When applying *C. minitans* follow the general principles highlighted below:

- Determine the volume of water needed to provide thorough coverage of the treatment area.
- Partially fill the spray tank with clean water and begin agitation. Add the specified amount of product to the tank and fill the tank to the volume needed to provide maximum coverage (refer to product label).
- Maintain agitation throughout spray application. Do not allow the spray mixture to stand overnight or for prolonged periods.
- Apply directly to the soil surface, plant parts and plant debris that will eventually have contact with the soil surface.
- Avoid application in saturated or waterlogged soil.
- Irrigate soil immediately after application so that the product is incorporated into the top two to five cm of soil.
- A follow-up treatment of product is required if the upper layer of treated soil is disturbed after application due to crop thinning or cultivation, as this will bring untreated sclerotia from the lower layers to the top soil layer.
- Use *C. minitans* repeatedly to increase long-term reduction of the pathogen (follow specific label advice).

Products should be stored in a cool, dry location and used within the recommended time-frame (check product label for details of storage conditions and expiry date).

4.2.4 *Bacillus pumilus* for management of downy mildew, powdery mildew and other diseases of tobacco

BACKGROUND

- **Active substance:** *Bacillus pumilus*
- **Substance group:** Microbial – Bacteria
- **Target pests:** For use against a wide range of fungal diseases, including blue mould

- **Application site and/or stage:** Foliar application
- **Special considerations for use:** Apply in the evening or early in the morning. Do not apply in full sun. As with all biological agents wear appropriate PPE, as indicated on the product label.
- **Overview of the regulatory status:** *Bacillus pumilus* is registered for use in tobacco in many countries.

Bacillus pumilus is a common bacterium that occurs in soils and water worldwide. It is a broad-spectrum preventive biopesticide used for the control or suppression of many important diseases of tobacco, including *Pythium* spp., *Fusarium* spp., *Rhizoctonia* spp., *Alternaria* spp., *Aspergillus* spp. and blue mould (*Peronospora hyoscyami*). It acts as a fungicide by forming a physical barrier between the plant surface and the fungal spores, inhibiting fungal development on the plant surface and then colonizing the fungal spores. It also stimulates treated plants' immune systems. It promotes root growth as well, which helps in the development of vigorous root systems and uniform plants. It is not harmful to human health or the environment.

Multiple strains of *B. pumilus* have been commercialized and are allowed for use in tobacco. Peat-based products containing *B. pumilus* are available, which can be used as a growing medium for tobacco seedlings for transplants. This can be used in floating seedbed systems. Other products containing *B. pumilus* can be applied with normal spray equipment for ground application or chemigation. It can be applied to the soil for management of many diseases and also to the plant surface for management of blue mould.

Bacillus pumilus is most effective when used as part of a preventive disease management programme. It is compatible with many fungicides, bactericides, insecticides and adjuvants. It can be applied on its own, tank-mixed with other CPAs or used in rotation with other fungicides for improved management. Using *B. pumilus* with a surfactant can improve penetration and coverage of above-ground portions of the tobacco plants. When conditions are conducive to heavy disease pressure, *B. pumilus* should be used in a rotational programme with other registered fungicides. Likewise, when the weather conditions are expected to be conducive to disease development, if the field has a history of disease problems, or if minimum/low-till programs are in place, use the higher application rates suggested.

Bacillus pumilus can be applied up to and including the day of harvest. It should be stored in a dry place at 25°C.

4.2.5 *Bacillus subtilis* for management of plant pathogens

BACKGROUND

- **Active substance:** *Bacillus subtilis*
- **Substance group:** Microbial – Bacteria
- **Target pests:** Oomycetes, Fungi

- **Application site and/or stage:** For plant disease control, these include foliar application and products applied to the root zone, compost, or seed.
- **Special considerations for use:** While *B. subtilis* is not a known human pathogen or disease-causing agent, it does produce the enzyme subtilisin, which has been reported to cause dermal allergic or hypersensitivity reactions in individuals repeatedly exposed to this enzyme in industrial settings. As with all biological agents wear appropriate PPE, as indicated on product label.
- **Overview of the regulatory status:** *Bacillus subtilis* is registered for use in tobacco in many countries.

Bacillus subtilis is a ubiquitous, naturally occurring, saprophytic bacterium that commonly occurs in soil, water, air and decomposing plant material. Under most conditions, however, it is not biologically active and is present only in spore form. Different strains of *B. subtilis* can be used as biological control agents in different situations. There are two general categories of *B. subtilis* strains: those that are applied to the foliage of a plant and those that are applied to the soil or transplant mix when sowing or transplanting. *Bacillus subtilis* bacteria act through:

- Antibiosis: *Bacillus subtilis* produces natural antibiotics that enable it to out-compete other microorganisms either by killing them or by reducing their growth rate.
- Competition for sites and/or nutrients.
- Induction of systemic acquired resistance (SAR).
- Improvement of nutrient utilization; this leads to overall increased plant health, which may improve crop vigour, yields and quality, especially under stressful conditions.

When applying *B. subtilis* as a spray follow the general principles indicated below:

- Adjust the application rate and/or spray intervals of *B. subtilis* according to the product label instructions.
- Ensure that product use conforms to resistance management strategies, which may include rotating and/or tank mixing with other products with different modes of action.
- Use the product in sufficient water to achieve thorough coverage (see label instructions).
- Maintain agitation during mixing and application to ensure uniform product suspension.
- Add a surfactant, known to be safe for tobacco, to the spray tank to improve penetration and coverage of above-ground portions of the plant.
- Use an appropriate application rate and/or spray intervals (refer to product label).
- Conduct a spray compatibility test if a mixture with other pesticides, surfactants, or fertilizers is planned.
- Do not exceed label dosage rates.

- Maintain a spray solution with a pH between 4.5 and 8.
- Heavy rainfall or irrigation shortly after application may result in the need to re-apply.

  Products should be stored in a cool, dry location and used within rec-ommended time-frame (check product label for details of storage condi-tions and expiry date).

4.2.6 Tea tree oil for management of plant pathogens in tobacco

BACKGROUND

- **Active substance:** Extract of *Melaleuca alternifolia*
- **Substance group:** Botanical
- **Target pests:** powdery mildew, downy mildew, early and late blight, *Botrytis*, *Sclerotinia*, *Fusarium*, *Rhizoctonia*, *Cladosporium* and *Cercospora*
- **Application site and/or stage:** For plant disease prevention and control by foliar application.
- **Special considerations for use:** As with all biological agents wear ap-propriate PPE, as indicated on product label.
- **Overview of the regulatory status:** Products containing tea tree oil ex-tracts are registered for use in tobacco in several countries.

Tea tree oil is extracted from the leaves of the myrtle tree (*Melaleuca al-ternifolia*). It contains compounds that can help to prevent and control a wide range of pathogens in tobacco. These include powdery mildew, downy mildew, early and late blight, *Botrytis*, *Sclerotinia*, *Fusarium*, *Rhizoctonia*, *Cladosporium* and *Cercospora*. Tea tree oil is available in commercial formulations.

Tea tree oil is used to control plant pathogens by affecting the pathogen cell structure. It is unlikely to have adverse effects on non-target organisms because it has a relatively low toxicity and tends to evaporate or degrade rapidly after application. However, products containing tea tree oil can be moderately irritating to skin and eyes, and are a skin sensitizer.

Guidelines for use:

- Apply tea tree oil in the early stages of infection.
- Do not apply when temperatures exceed 35°C, or during windy condi-tions in the field.
- Dilute the tea tree formulation as recommended on the label. Do not exceed 0.7 kg tea tree oil per acre.
- Apply tea tree oil in the greenhouse, nursery or the field using spray equipment.
- Ensure thorough coverage of the plant with the product but do not apply to the point of excessive runoff.
- Observe a restricted entry interval of 4 h.

- Apply at 1–2-week intervals throughout the growing season. If disease pressure is high, use the shorter spray interval.
- It is advisable to obey a 48 h pre-harvest interval.

When using tea tree oil, wear chemical-resistant goggles, respirator, long-sleeved shirt and long pants, chemical-resistant gloves and shoes plus socks.

4.3   Managing Nematodes with Biopesticides

4.3.1   *Paecilomyces lilacinus* for nematode management

BACKGROUND

- **Active substance:** *Paecilomyces lilacinus* (current taxonomic designation: *Purpureocillium lilacinus*)
- **Substance group:** Microbial – Fungus
- **Target pests:** Nematodes
- **Application site and/or stage:** Field
- **Special considerations for use:** *Paecilomyces lilacinus* can cause rare opportunistic infections in humans, with significant resistance to conventional antifungals. As with all biological agents wear appropriate PPE, as indicated on product label.
- **Overview of the regulatory status:** *Paecilomyces lilacinus* is registered for nematode management in some countries where tobacco is grown. Check the national list of registered pesticides to confirm that it is registered and allowed for use in tobacco.

Paecilomyces lilacinus is a naturally occurring fungus commonly found in soils globally. It can potentially control a range of plant parasitic nematodes, including those from the genera *Meloidogyne*, *Pratylenchus* and *Radopholus*. This nematophagous fungus acts through its hyphae by directly parasitizing the nematode eggs and the emerging larvae. It does not affect beneficial insects.

Paecilomyces lilacinus is produced as conidia in water-dispersible granules. It is generally applied as a seedling or soil drench at the base of the plant, but it can also be applied by spray application or drip irrigation (refer to product label for specific application timing and concentrations). An example of its use would be to apply a number of days before planting, at planting and then at recommended periods post-planting/transplant.

Repeated application to maintain the antagonist population at a sufficient level could be used to secure long-term control of pathogenic nematodes.

To increase its efficacy, treatment with *P. lilacinus* can be combined with soil solarization or other soil-applied biocontrol agents in the field.

Products should be stored in a cool, dry location and used within recommended time-frame (check product label for details of storage conditions and expiry date).

4.3.2 *Bacillus firmus* for nematode management

BACKGROUND

- **Active substance:** *Bacillus firmus*
- **Substance group:** Microbial – Bacteria
- **Target pests:** Nematodes, primarily *Meloidogyne* spp. but also other nematodes such as *Trichodorus* spp., *Ditylenchus* spp.
- **Application site and/or stage:** Depending on the product, *B. firmus* may be applied as a seed treatment, in the seedbed or in the field.
- **Special considerations for use:** Apply during late evening to reduce temperature effects on its establishment. As with all biological agents wear appropriate PPE, as indicated on product label.
- **Overview of the regulatory status:** *Bacillus firmus* is registered for nematode management in some countries where tobacco is grown. Check the registered pesticide list to confirm that it is registered and allowed for use in tobacco.

Bacillus firmus is a naturally occurring soil bacterium that associates with the root systems of plants and possesses nematicidal activity. *B. firmus* colonizes the egg sacs of rootknot nematodes, subsequently destroying the nematode eggs. Thus it serves to suppress nematode populations, providing early season protection against nematodes. Treatment with *B. firmus* results in improved plant vigour for more uniform, higher yielding plants. Studies in crops such as tomato have demonstrated that it can achieve control levels equivalent to those of chemical controls and can keep nematode populations below economic threshold levels. In areas with high nematode infestation additional control measures may be needed. There is no evidence that *B. firmus* poses a risk to human health, and it is non-harmful to non-target organisms and the environment.

Commercialized products containing *B. firmus* are applied as a seed treatment or directly to the soil in the greenhouse or field. For products that are used as seed treatments, apply *B. firmus* as a water-based slurry. When used in the field, it is generally recommended that the product is applied several days prior to planting to treat the soil by spraying or by drip irrigation in furrows. It can also be applied for a second time mid-season for additional control.

Bacillus firmus is compatible with many seed treatments and soil fungicides. It is not compatible for use at the same time as soil sterilants, but it can be used after the application of soil sterilants for improved control. *Bacillus firmus* is thermo-tolerant, so it can be used in conjunction with soil solarization. For more information on how to use products containing *B. firmus,* refer to the product's label.

Products containing *B. firmus* should be stored in a sealed container under cool, dry conditions (check product label for details of storage conditions and expiry date).

4.4   Managing Invertebrate Pests of Tobacco with Biopesticides

4.4.1   Using *Metarhizium* to manage grasshoppers and other insect pests

BACKGROUND

- **Active substance:** *Metarhizium* spp.
- **Substance group:** Microbial – Fungus
- **Target pests:** Many significant pests of tobacco including caterpillars, beetles, mites, whiteflies, aphids and thrips
- **Application site and/or stage:** Field
- **Special considerations for use:** Apply in the evening or early in the morning. Do not apply in full sun as it is UV sensitive. As with all biological agents wear appropriate PPE, as indicated on product label.
- **Overview of the regulatory status:** *Metarhizium* spp. are registered for use in tobacco in many countries. In Africa in particular it has been used successfully to replace many highly hazardous pesticides for the control of locusts and other swarming grasshoppers.

Metarhizium is an entomopathogenic fungus, i.e. a type of fungus that parasitizes insects. It can kill or disable the target insect host. *Metarhizium* spp. are found in soil or as endophytes (beneficial plant symbionts). The spores or mycelium can attach to the surface of the insects. They can penetrate the host insect's cuticle and use enzymes to penetrate the insect. Once inside the host insect body cavity *Metarhizium* can multiply. Once the insect is dead, the fungus colonizes the insect and then sporulates. These spores are then disseminated and the process begins again. The application of these entomopathogens to control insect pests in the field is well established and many commercial products are available.

The products are composed of dry spores which can be mixed with agricultural adjuvants or sprayed in oil formulations.

Metarhizium products should be stored in a cool, dry location and used within the recommended time-frame (check product label for details of storage conditions and expiry date).

Topics covered below include:

- Recogniziing *Metarhizium*
- Understanding what entomopathogens are
- How *Metarhizium* can be used to kill insects

4.4.1.1   *ACTIVITY: To illustrate the form of Metarhizium to trainees*

Time required: Initially 30 min. Subsequent observation over 5 days (or pre-prepare as visual demonstration)
Target audience: Farmers and field technicians
Location: Classroom

MATERIALS:

Microscope
Pre-prepared slides
Photo sheets illustrating conidiophores and spores of *Metarhizium*

If a laboratory and materials are available the following could also be pre-pared to provide further information to the trainees.

Prepare slides of *Metarhizium*.

Time required: 5–8 days

MATERIALS:

- Microscope
- Dissecting needles
- Alcohol
- Petri dishes
- Potato dextrose agar (commercial or home-made)
- Distilled water
- Microscope slides and cover slip
- Cotton blue stain
- Burner

PROCEDURES:

- Prepare 5 cm or 9 cm agar plates containing 10% potato dextrose agar or Sabouraud agar.
- Inoculate one edge of the plate with *Metarhizium* spores.
- Incubate at 25°C in the dark for 4–7 days.
- When the *Metarhizium* has colonized two-thirds of the plate and is sporulating, remove a small amount of the mycelia and place into a drop of cotton-blue stain. Tease the mycelia apart and place a cover slip over them.
- View under the microscope to identify conidiophore and conidia.
- If the slide is useful for demonstration then seal the edge with nail varnish.
- Prepare a number of slides for demonstration.

4.4.1.2   ACTIVITY: Understanding what entomopathogenic fungi are

Time required: Initially 30 min. Subsequent observation over 14 days (or pre-prepare as visual demonstration)
Target audience: Farmers and field technicians
Location: Classroom

MATERIALS:

- Commercial or local *Metarhizium* product (spores)
- Field-collected insects adults/larvae
- Handheld sprayer

- Vegetable oil
- 70% alcohol
- Clean containers
- Non-absorbent cotton wool
- Printed table
- Flipchart

PROCEDURES:

Often farmers or other technical staff may not appreciate that biocontrol agents can be effective replacements for CPAs. This training topic looks at entomopathogenic fungi such as *Metarhizium*. It will provide trainees with an understanding of what entomopathogenic fungi are and the benefits their use can provide as part of an IPM programme.

Prepare flipchart or poster illustrating key points about entomopathogens including *Metarhizium*. Discuss with trainees their knowledge and experience of using entomopathogens.

- Split into groups.
- Use an alternative host such as *Galleria melonella*, sourced commercially, or field-collected insect larvae or adults.
- Depending on the number of available insects, place 10 insects in a clean container. Use three replicates for each treatment.
- Prepare spore inoculum in vegetable oil (4×10^7 spores/ml).
- Spray or apply drops of spore inoculum directly to each set of 10 insects in sterile containers, using three replicates.
- Repeat the process using uninoculated oil as a control.
- Incubate at ambient temperature and assess mortality over 14 days.
- Calculate the percentage mortality of the treatment as follows:

  No. killed by spore treatment/no. of insects used × 100

- Compare with control.

	No. dead insects	% mortality
Spore treatment		
Oil control		
% change		

QUESTION FOR DISCUSSION

- What was observed and what could this mean for control of pests?

Note. Alternatively facilitators can set up demonstration materials by preparing treatments with inoculations with entomopathogens and the control treatments.

This activity could also be carried using the insect zoo format (see below).

4.4.1.3   ACTIVITY: Insect zoo to illustrate use of entomopathogens

Time required: 2 h. Subsequent observation over 5 days (or pre-prepare as a visual demonstration)
Target audience: Farmers and field technicians
Location: Field and classroom

MATERIALS:

- Commercial or local *Metarhizium* product (spores)
- Field-collected insects; adults and/or larvae
- Handheld sprayer
- Vegetable oil
- 70% alcohol
- Non-absorbent cotton wool
- Small plastic vials or empty water containers
- Plastic bags
- Plastic buckets (transparent if available), large enough to hold tobacco leaves and plant parts of various sizes
- Tobacco leaves, stems, etc.
- Tissue paper
- Camel or fine hair brush
- Labels
- Muslin cloth or fine mesh screen
- Rubber bands/pieces of string
- Hand lens
- Printed table

PROCEDURES:

Very early in the morning, participants should carefully collect both unknown and known insects from the field plot using a sweep net or by capturing them in plastic vials/bottles. Be careful when handling the insects to be studied, as they will not feed if they have been roughly handled. Ask participants to study the insects and give the local name of each.

To set up the insect zoos, line the plastic buckets with tissue paper to avoid condensation. Put some leaves or other plant parts in each bucket and label each bucket with the (local) name of the insect to be studied. Put different insect species in different zoos.

- Depending on the number of insects available, place 10 insects in a clean container.
- Prepare spore inoculum in vegetable oil (4×10^7 spores/ml).
- Spray or apply drops of spore inoculum directly to one set of 10 insects, in a sterile container. Use three replicates for each treatment.
- Repeat the process using oil as a control.

- Maintain the containers at ambient temperature and assess mortality over 14 days.
- Calculate percentage mortality of the treatment as follows:

    No. killed by spore treatment/no. of insects used × 100

- Compare with control.

	No. dead insects	% mortality
Spore treatment		
Oil control		
% change		

At the end of the exercise, participants should present their observations to the wider group. Findings should be recorded on poster paper, including the following:

- Local names of the insect(s) collected
- Where the insects were collected
- What they fed on
- Whether they were killed by the entomopathogen

QUESTIONS FOR DISCUSSION

- What was observed and what could this mean for the control of pests?
- Are there any issues for non-target insects?

4.4.2 *Beauveria bassiana* for managing insect pests in tobacco

BACKGROUND

- **Active substance:** *Beauveria bassiana*
- **Substance group:** Microbial – Entomopathogenic fungus
- **Target pests:** Many significant pests of tobacco, including caterpillars, beetles, mites, whiteflies, aphids and thrips
- **Application site and/or stage:** Field
- **Special considerations for use:** Apply in the evening or early in the morning. Do not apply in full sun. As with all biological agents wear appropriate PPE, as indicated on product label.
- **Overview of the regulatory status:** *Beauveria bassiana* is one of the most widely registered biopesticides for use in tobacco globally.

Beauveria bassiana is one of the biopesticides most commonly used worldwide to control arthropod species. It occurs naturally in the soil throughout the world and is registered for use in many countries. There are many potential advantages to using *B. bassiana*; for example, residues are not an issue, and it can be applied up to the day of harvest. Depending on the circumstances in which it is used, the cost and labour requirements for *B. bassiana* may be similar to those of other CPAs. It is a reduced-risk CPA that poses little hazard to human health or the environment. When *B. bassiana* is applied regularly in combination with other cultural controls, farmers are often able to replace or reduce their use of other CPAs. Since *B. bassiana* is a biopesticide containing living spores, its use differs in some ways from that of other chemical CPAs. If not used correctly, it may be ineffective.

Beauveria bassiana is an entomopathogenic fungus, and has a broad host range. It attacks many significant pests of tobacco, including caterpillars, beetles, mites, whiteflies, aphids and thrips. *Beauveria bassiana* can be applied to the soil or as a foliar application. When an insect host comes into contact with *B. bassiana* spores, the spores stick to the insect's skin (cuticle), allowing the fungus to infect and kill the insect.

Beauveria products should be stored in a cool, dry location and used within the recommended time-frame (check product label for details of storage conditions and expiry date).

Topics covered below include:

- Recognizing *Beauveria*
- Understanding what entomopathogens are
- Understanding how *Beauveria* can be used to kill insects

4.4.2.1 *ACTIVITY: To illustrate the form of* Beauveria *to trainees*

Time required: Initially 30 min, with subsequent observations over 14 days (or pre-prepare as a visual demonstration)
Target audience: Farmers and field technicians
Location: Classroom

MATERIALS:

Microscope
Pre-prepared slides
Photo sheets illustrating conidiophores and spores of *Beauveria*

Prepare slides of Beauveria

If a laboratory and materials are available the following could also be prepared to provide further information to the trainees.
Time required: 5–8 days

MATERIALS:

- Microscope
- Dissecting needles
- Alcohol
- Petri dishes
- Potato dextrose agar (commercial or home-made)
- Distilled water

- Microscope slides and cover slip
- Cotton blue stain
- Burner

PROCEDURES:

- Prepare 5 cm or 9 cm agar plates containing 10% potato dextrose agar or Sabouraud agar.
- Inoculate one edge of the plate with *Beauveria* spores.
- Incubate at 25°C in the dark for 4–7 days.
- When the *Beauveria* has colonized two-thirds of the plate and is sporulating, remove a small amount of mycelia and place into a drop of cotton-blue stain. Tease the mycelia apart and place a cover slip over them.
- View under the microscope to identify conidiophore and conidia.
- If the slide is useful for demonstration then seal the edge with nail varnish.
- Prepare a number of slides for demonstration.

4.4.2.2   *ACTIVITY: Understanding what* Beauveria bassiana *is*

Time required: Initially 30 min. Subsequent observation over 14 days (or pre-prepare as visual demonstration)
Target audience: Farmers and field technicians
Location: Classroom

MATERIALS:

- Commercial or local *Beauveria* product (spores)
- Field-collected insects; adults and/or larvae
- Handheld sprayer
- Vegetable oil
- 70% alcohol
- Clean containers
- Non-absorbent cotton wool
- Printed table
- Flipchart

PROCEDURES:

Farmers or other technical staff may often not appreciate that biocontrol agents can be effective replacements for CPAs. This training topic looks at entomopathogenic fungi such as *Beauveria*. This training activity will provide trainees with an understanding of what entomopathogenic fungi are and the benefits their use can provide as part of an IPM programme.

Prepare a flipchart or poster illustrating the key points about entomopathogens, including *Beauveria*. Discuss with trainees their knowledge and experience of using entomopathogens.

- Split into groups.
- Use an alternative host such as *Galleria melonella*, sourced commercially, or field-collected insect larvae or adults.
- Depending on the number of insects available, place 10 insects in a clean container, using three replicates for each treatment.
- Prepare spore inoculum in vegetable oil (4×10^7 spores/ml).
- Spray or apply drops of spore inoculum directly to each set of 10 insects, in sterile containers.
- Repeat the process using oil as a control.
- Incubate at ambient temperature and assess mortality over 14 days.
- Calculate percentage mortality of the treatment:

  No. killed by spore treatment/no. of insects used $\times$ 100

- Compare with control.

	No. dead insects	% mortality
Spore treatment		
Oil control		
% change		

QUESTION FOR DISCUSSION

- What did the participants observe and what could this mean for the control of pests?

Note. Facilitators: alternatively, set up demonstration materials by preparing inoculations with entomopathogens, together with controls.

This activity could also be carried using the insect zoo format (see below).

4.4.2.3   *ACTIVITY: Insect zoo to illustrate use of entomopathogens*

Time required: 2 h. Subsequent observation over 5 days (or pre-prepare as visual demonstration)
Target audience: Farmers and field technicians
Location: Field and classroom

MATERIALS:

- Commercial or local *Beauveria* product (spores)
- Field collected insects; adults and/or larvae
- Handheld sprayer
- Vegetable oil
- 70% alcohol
- Non-absorbent cotton wool

- Small plastic vials or empty water bottles containers
- Plastic bags
- Plastic buckets (transparent if available), large enough to hold tobacco leaves and plant parts of various sizes
- Tobacco leaves, stems, etc.
- Tissue paper
- Camel- or fine-hair brush
- Labels
- Muslin cloth or fine mesh screen
- Rubber bands/pieces of string
- Hand lens
- Printed table

PROCEDURES:

Very early in the morning, participants should carefully collect both known and unknown insects from the field plot using a sweep net or by capturing them in plastic vials/bottles. Be careful when handling the insects to be studied, as they will not feed if they have been roughly handled. Ask participants to study the insects and give the local name for each one.

To set up the insect zoos, line the plastic buckets with tissue paper to avoid condensation. Put some leaves or other plant parts in each bucket and label each bucket with the (local) name of the insect to be studied. Put different insect species in different zoos.

- Depending on the number of insects available, place 10 insects in a clean container.
- Prepare spore inoculum in vegetable oil (4×10^7 spores/ml).
- Spray or apply drops of spore inoculum directly to one set of 10 insects in a sterile container.
- Repeat the process using oil as a control.
- Maintain the containers at an ambient temperature and assess mortality over 14 days
- Calculate percentage mortality of the treatment as follows:

  No. killed by spore treatment/no. of insects used × 100

- Compare with control.

	No. dead insects	% mortality
Spore treatment		
Oil control		
% change		

At the end of the exercise, participants should present their observations to the wider group. Findings should be recorded on poster paper, including the following:

- Local names of the insect(s) collected
- Where the insects were collected
- What they fed on
- Whether they were killed by the entomopathogen

QUESTIONS FOR DISCUSSION

- What did the participants observe and what could this mean for the control of pests?
- Are there any issues for non-target insects?

4.4.2.4 ACTIVITY: Demonstrate how to use Beauveria *spp. in the field*

Time required: 1 h
Target audience: Farmers
Location: Infested field of tobacco

MATERIALS:

- Product containing *Beauveria* spp.
- Spray application equipment
- Appropriate PPE

PROCEDURES:

Begin by explaining that *Beauveria* will be most effective if it is applied when pests are detected at the onset of infestation.

Emphasize that early scouting and detection followed by application of *Beauveria* when insect numbers are low will result in the most effective control.

In the demonstration, make note of the following key points regarding the mode of application and application timing:

- Wear appropriate PPE as described on the label when applying and mixing.
- Always check the product label for dosage and target pests. Water volume depends on spray equipment, crop canopy and target pest.
- Apply in the evening or early in the morning. Do not apply in full sun.
- Shake the container well before mixing to ensure that spores are suspended.
- To mix, fill spray tank with half the desired amount of water and agitate to mix. Shake the product, slowly add the desired quantity to the spray tank and then add the remaining amount of water.
- Once mixed, *Beauveria* will only be viable for a limited time, so the mixture should be applied as soon as possible. Do not mix more product than is needed for that day.
- Continue agitation during spraying.
- Apply as recommended on the product label.

- *Beauveria* may take some days to infect and kill pests. When immediate control is required in order to avoid crop losses, pair it with an appropriate insecticide that is compatible for use with *Beauveria*.
- Many fungicides are not compatible with *Beauveria*. Do not tank mix *Beauveria* with fungicides unless compatibility can be assured. Wait at least 48 h before applying fungicides to areas that have been treated with *Beauveria*.
- Avoid applying to areas where honey bees are active.
- There is no limit on the number of applications or the total amount of product that can be applied in one season.
- There is no preharvest interval for *Beauveria*. It can be applied up to the day of harvest.

Since *Beauveria* is a living organism, storage conditions are important. Make note of the following points:

- Always store the product in a cool, dry place out of direct sunlight (between 40°F/4°C and 80°F/27°C). Avoid storing or transporting at high temperatures as this may harm the live spores of *Beauveria*.
- Close containers tightly after use in order to avoid contamination and loss of efficacy. Whenever possible, once a container has been opened, use all of its contents.
- Do not store or let *Beauveria* preparations stand for more than 4 h.

4.4.3   Neem-based products for managing whiteflies, aphids, thrips and caterpillars in tobacco

BACKGROUND

- **Active substance:** Azadirachtin (and other limonoids/terpenoids)
- **Substance group:** Plant extract/Botanical
- **Target pests:** Insects
- **Application site and/or stage:** Soil or foliar application, field
- **Special considerations for use:** Ensure appropriate PPE is worn, as indicated on the product label.
- **Overview of the regulatory status:** Registered for use in tobacco in many countries.

Neem extracts can control numerous insect pests, including several important ones found on tobacco such as whiteflies, aphids, thrips, caterpillars and leafhoppers. Neem products have also been shown to be effective against some pests that have developed resistance and are difficult to control by most synthetic insecticides (Lokanadhan *et al.*, 2012).

Most commercial neem products are oil-based, emulsifiable concentrates containing azadirachtin at various concentrations. Azadirachtin disrupts or inhibits the normal development of eggs, larvae or pupae by preventing normal hormone releases triggering growth and maturation. It also acts as a repellant, feeding and ovipositional deterrent, and mating disruptor (Ghewande et al., 1993).

Neem products are easy to use and relatively safe in comparison with conventional synthetic pesticides. In particular, neem is generally of low toxicity to humans, and does not persist in the environment.

Neem can be applied as a foliar spray application (see product label for application rates and timing). It has a slow mode of action, and its effects are usually not immediately visible. Some neem extracts (e.g. oil extracts) may be phytotoxic. Therefore, it is important to test the extract on plants before full-scale spraying in the field.

4.4.4 *Bacillus thuringiensis* for managing caterpillars and larvae (Lepidoptera) in tobacco

BACKGROUND

- **Active substance:** Bt: *Bacillus thuringiensis* subsp. *aizawi* or *B. thuringiensis* subsp. *kurstaki*
- **Substance group:** Microbial -- Bacteria
- **Target pests:** Caterpillars and larvae such as *Helicoverpa armigera, Helicoverpa zea, Spodoptera exigua, Spodoptera littoralis, Heliothis virescens*
- **Application site and/or stage:** Field
- **Special considerations for use:** As with all biological agents appropriate PPE should be worn, as indicated on product label.
- **Overview of the regulatory status:** Bt is one of the most widely registered biopesticides for use in tobacco globally.

Bacillus thuringiensis (*Bt*) is one of the most commonly registered biopesticides for use in tobacco around the world. *Bt* is a bacterium naturally found in soils worldwide and infects many insect pests, including beetles, mosquitoes, blackflies, caterpillars and moths. The proteins produced by the bacteria are toxic to insects and their action is very specific; each type of *Bt* strain targets a specific group of insects. *B. thuringiensis* subsp. *aizawi* and *B. thuringiensis* subsp. *kurstaki* infect lepidopteran larvae such as *Helicoverpa armigera, Helicoverpa zea, Spodoptera exigua, Spodoptera littoralis, Heliothis virescens*. For this reason the type of *Bt* to be used must be selected carefully to match and control the target insect pest. To be effective, the bacterium must be eaten by caterpillars and reach the gut of the insect before it will take effect. When small caterpillars and larvae start feeding on tobacco leaves sprayed with *Bt*, the bacteria reach the gut where the toxins are activated and start to break down the gut. The infected larva dies of infection and starvation.

As *Bt* strains are highly specific to insect groups, *Bt* sprays usually do not harm beneficial insects. Other benefits of using *Bt* include no residue issues, and it is safe for use in the environment, with no known effects on wildlife. As some insect pests have developed resistance due to intensive spraying over many years, a pesticide resistance management strategy is highly recommended.

How to use:

- Apply when eggs are due to hatch or when caterpillars or larvae are small (1st and 2nd instar larvae). These are the most susceptible instars, because the larvae feed on open leaf surfaces that are accessible to sprays. Once the larvae have bored into the stem, they are protected from sprays.
- Early scouting and detection followed by application of *Bt* products when caterpillar and larvae numbers are low will result in the most effective control.
- Wear appropriate PPE as described on the label when applying and mixing.
- Check the product label for dosage and target pests. Water volume depends on spray equipment, crop canopy and target pest.
- *Bt* breaks down under the UV light of the sun, therefore apply in the evening or early in the morning. Do not apply in full sun.
- Do not apply *Bt* product through any type of irrigation system.
- To mix, fill spray or mixing tank three quarters full with the desired amount of water and start agitation. Slowly pour the desired quantity of product into water with the agitator running, and add the remaining amount of water. Agitate as necessary to maintain suspension.
- Continue agitation during spraying.
- Additional adjuvants, spreaders or stickers approved for tobacco may be added to improve product performance, especially under heavy dew or rainy conditions.
- Use diluted sprays as soon as possible and within 48 h. Do not mix more product than is needed for that day.
- For adequate insect control, thorough and uniform crop coverage is required. Ensure good spray coverage on the top and bottom of the foliage.
- To assure efficacy of the product, larvae must be actively feeding on treated, exposed plant surfaces.
- There is no preharvest interval for *Bt*. It can be applied up to and on the day of harvest.
- Re-entry interval (REI) is 4 h before entering treated areas.
- Repeat applications as necessary under a pest management programme that includes close scouting, usually at 3- to 14-day intervals depending on plant growth rate, moth egg-laying activity and rainfall after treatment.
- In the case of a heavy infestation, where quick action is required to avoid crop losses, pair with an appropriate contact insecticide compatible for use with *Bt*. *Bt* products can be mixed with commonly used insecticides or fungicides, which are generally not deleterious to the *Bt* product if the mix is used promptly. Before mixing in the spray tank, it is recommended to test physical compatibility by mixing all components in a small container in proportionate quantities.

- *Bt* is toxic to aquatic invertebrates and highly toxic to honey bees exposed to direct treatment. Do not apply directly to water or while bees are actively visiting the treatment area.

Bt should be stored in its original container in a cool, dry place (under 90°F/32°C), inaccessible to children and away from heat and direct sunlight. Protect from freezing.

4.4.5 *Trichogramma* spp. for managing lepidopteran pests

BACKGROUND

- **Active substance:** *Trichogramma* spp., e.g. *T. pretiosum* or *T. evanescens*
- **Substance group:** Macroorganism – Parasitoid wasp
- **Target pests:** Caterpillars and larvae such as *Helicoverpa* spp., *Heliothis virescens*, *Ephestia elutella* (in storage)
- **Application site and/or stage:** Field/Storage
- **Special considerations for use:** Do not use on small, isolated fields
- **Overview of the regulatory status:** With a few notable exceptions (e.g. Brazil), most countries do not require native *Trichogramma* spp. to be registered, but be careful when considering the use of exotic *Trichogramma* spp.

Trichogramma is one of the most widely used biological control agents. Despite its tiny size — 1 mm or less — *Trichogramma* is an efficient natural enemy of lepidopteran pests of tobacco such as budworm and tobacco moth. *Trichogramma* are released as parasitized eggs attached to a piece of card known as a Tricho-card. Adult *Trichogramma* females emerging from these Tricho-cards search for pest eggs into which they can lay their eggs. This will kill the larvae before they hatch, thus preventing them from doing any damage. The release of *Trichogramma* can have many advantages. For example, residues are not an issue, and Tricho-cards can be applied up to the day of harvest. Also, applying Tricho-cards poses no hazard to human health or the environment. Applying *Trichogramma* in combination with other cultural controls means that farmers are often able to replace or reduce their use of conventional CPAs.

4.4.5.1 *ACTIVITY: Demonstrate how to use* Trichogramma *wasps in the field*

Time required: 1 h
Target audience: Farmers
Location: Infested field of tobacco

MATERIALS:

- Tricho-cards
- Parasitized eggs and eggs that have not been parasitized (or photos of parasitized eggs and eggs that have not been parasitized)

PROCEDURES:

Demonstrate to the participants the parasitized eggs and the eggs that have not been parasitized (or photographs of these). Explain that *Trichogramma* wasps are very small but also very effective, killing the moth eggs before they can hatch and cause any damage.

Instruct the participants to follow these steps in order to be able to use *Trichogramma* wasps for the management of lepidopteran pests in tobacco fields:

- Place the Tricho-cards into the field shortly before the *Trichogramma* wasps emerge from the eggs on the card (based on information from the provider).
- Make sure that the printed side is facing up as this will protect the *Trichogramma* on the back side from rain and direct sunlight.
- Avoid release during hot times of the day and during or before heavy rain.
- Put out 100 Tricho-cards for each hectare of field. The release points should be about 10 m apart. Each Tricho-card harbours about 1000 *Trichogramma* wasps, resulting in the release of 100,000 wasps per hectare.
- The *Trichogramma* wasps emerging from the cards could be killed by any insecticides applied to the crop. Thus, DO NOT spray pesticides, particularly insecticides, either 1 week before or after the release of *Trichogramma*!
- With *Trichogramma* releases, insecticide applications are often made unnecessarily. If there happens to be serious pest occurrence, consult a local plant protection agency for advice.
- *Trichogramma*, and many other beneficial insects, can benefit from nectar sources around the treated crop. It would therefore be an advantage to preserve and/or grow plants providing flowers and nectar, for example, sesame or soybean plants.
- The efficacy of *Trichogramma* is very much improved if they are released into a larger area, e.g. if farmers from a village agree to combine their efforts. If they are only released in a single field, some *Trichogramma* might leave that field, thereby reducing their effectiveness.

4.4.6 Nucleopolyhedrosis virus for managing caterpillars and larvae (Lepidoptera) in tobacco

BACKGROUND

- **Active substance:** Nucleopolyhedrosis virus (NPV) of *Angrapha falcifera*, *Autographa californica*, *Helicoverpa armigera*, *H. zea*, *Spodoptera exigua*, *S. littoralis*
- **Substance group:** Microbial -- Baculovirus

- **Target pests:** Caterpillars and larvae such as *Helicoverpa armigera*, *Helicoverpa zea*, *Spodoptera exigua*, *Spodoptera littoralis*, *Heliothis virescens*
- **Application site and/or stage:** Field
- **Special considerations for use:** As with all biological agents wear appropriate PPE, as indicated on product label. Do not expose to direct sunlight.
- **Overview of the regulatory status:** NPV is registered for use in tobacco in many countries globally.

Nucleopolyhedrosis virus (NPV) is highly specific, safe and environmentally friendly, making it ideally suited for inclusion in an integrated pest management (IPM) approach in cropping systems. NPV particles are called polyhedral inclusion bodies (PIBs) and must be eaten by the larvae for infection to occur. Once the PIB is ingested, the virus infects the epithelial cells of the midgut. Ingestion of a single PIB is usually sufficient to kill the insect. NPV belongs to a group of insect diseases called baculoviruses that infect and kill the larvae of moths and sawflies. NPV can kill young larvae within 4 days of ingestion, and older larvae within 5 to 7 days, depending on dose and temperature. NPV is self-propagating in the field. The infection of one caterpillar/larva will result in the death of a whole population. This reduces the number of chemical sprays generally required in the field.

NPV will not affect non-target organisms.

NPV is usually formulated as the lyophilized powder of dead insect larvae/caterpillars or a suspension of the PIB.

NPV can be applied in the field in the same way as a standard CPA.

NPV should be stored in a cool dry location Powder formulations may be stored frozen but not liquid formulations.

4.4.7 *Aphidius* spp. for managing aphids in tobacco

BACKGROUND

- **Active substance:** *Aphidius colemani*, *Aphidius ervi*, *Aphidius matricariae*
- **Substance group:** Macrobial – Parasitoid
- **Target pests:** Aphids – *Myzus persicae*, *Myzus persicae nicotianae*
- **Application site and/or stage:** Seedbed / Field
- **Special considerations for use:** *Aphidius* is less effective at temperature above 30°C
- **Overview of the regulatory status:** Verify if *Aphidius* spp. are permitted for release according to the national regulations.

Aphidius wasps are tiny black parasitic wasps that complete their development inside the aphid's body, which takes between 13 to 15 days. The female wasp lays a single egg within an aphid. The parasitized aphid continues to feed while the wasp larva develops and progresses through four stages. The wasp larva then pupates within the aphid's body. As a result, the aphid swells and its skin hardens and turns golden brown in colour, which is called 'mummy'. Once the wasp has completed its development, it makes a round hole in the mummy to exit and find new aphids to parasitize. One female *Aphidius* can lay between 100 to over 400 eggs in its lifespan, depending on the species. Aphidius wasps are efficient searchers and can detect aphid colonies from a long distance. They can be used on a wide range of crops, including horticultural, ornamental, small fruit crops and open field crops as well as in greenhouses. As aphid colonies often consist of several aphid species, some providers sell mixture of *Aphidius* species (*Aphidius colemani*, *Aphidius ervi*, *Aphidius matricariae*). *A. colemani* attacks a wider range of aphid species compared with *A. matricariae*, which is especially effective against the green peach aphid, *Myzus persicae*. All three species are very effective against aphids that are spread in small, well-dispersed populations.

When to apply:

- The rates, interval and frequency of release of *Aphidius* wasps depend on the climate conditions and the level of infestation. It is recommended to use *Aphidius* at early infestation stage, as *Aphidius* wasps are more effective in parasitizing small-sized aphids even if they are able to parasitize aphids at any stage.
- Depending on the product and *Aphidius* species used, recommended dose can vary:

  - For preventive measures: use 0.25 to 1 individual/m^2 at 7- or 14-day interval.
  - Curative light: one individual/m^2 at 7-day interval, minimum three times.
  - Curative heavy: 2/m^2 at 7-day interval, minimum six times.
  - Environmental conditions: *Aphidius* is less effective at temperatures above 30°C. Avoid releasing the wasps when temperatures are above 30°C.

How to use:

- Generally, *Aphidius* wasps are sold in bottles as mummies, which can be mixed with wood-chips or buckwheat.
- Before opening the bottle, set it horizontally and gently rotate it to mix the content.
- Use rockwool slabs to spread the material or alternatively spread the material in application boxes sold by the provider. Leave the material on the release sites for a few days, as the wasps can still emerge.
- When used as a preventive measure, hang the bottles on the plants or poles and distribute them over the crop homogeneously. Once the aphids are detected, concentrate the bottles/release boxes of parasitoids on aphid hot spots.
- Release the parasitoid wasps early in the morning or at sunset.

- Only use *Aphidius* if they are permitted for release in the country/region and crop. Check local registration requirements.
- The first mummies can be seen in the crop approximately 2 weeks after the first introduction.

Storage:

Aphidius species have a very short life expectancy and lay most of their eggs within the first 4 days. Therefore, it is important to introduce the wasps into the crops as soon as possible after receipt. Storage over a long period can have a negative impact on their quality and effectiveness to control aphids. If storage is needed, follow carefully the conditions below:

- Storage after receipt: 1–2 days maximum
- Storage temperature: 8–10°C/47–50°F
- Store in the dark, never expose it to sunlight or pesticides
- Keep the bottles horizontally in storage

4.4.8 Entomopathogenic nematodes for managing insect pests in tobacco

BACKGROUND

- **Active substance:** *Heterorhabditis* spp., *Steinernema* spp.
- **Substance group:** Microbial – nematodes
- **Target pests:** Entomopathogenic nematodes are effective against many insect pests of tobacco with soil dwelling stages of their lifecycle, including white grubs (*Phyllophaga* spp.), cutworms (*Agrotis* spp.) and armyworms (*Spodoptera* spp.)
- **Application site and/or stage:** Field
- **Special considerations for use:** Apply in the evening or early in the morning. Do not apply in full sun.
- **Overview of the regulatory status:** Most countries do not require the registration of native species of entomopathogenic nematodes.

Entomopathogenic nematodes (EPNs) are a group of nematodes that parasitize and kill many different types of insects that come into contact with the soil, such as armyworms, cutworms, white grubs, etc. The EPNs most commonly used for the biological control of insects are found within the genera *Steinernema* and *Heterorhabditis*. Rapid death by EPNs reflects the pathogenicity of their bacterial symbionts. In the case of *Steinernema* and *Heterorhabditis* these bacteria are from the *Xenorhabdus* and *Photorhabdus* genera, respectively. A simplified illustration of the infection and subsequent death of an insect host can be seen in Fig. 4.1 below.

EPNs employ different strategies to find host larvae; some species actively search for the host insect larvae while others wait and use an ambush strategy.

EPNs can be supplied as formulated products in clay or another carrier, or on moist sponge.

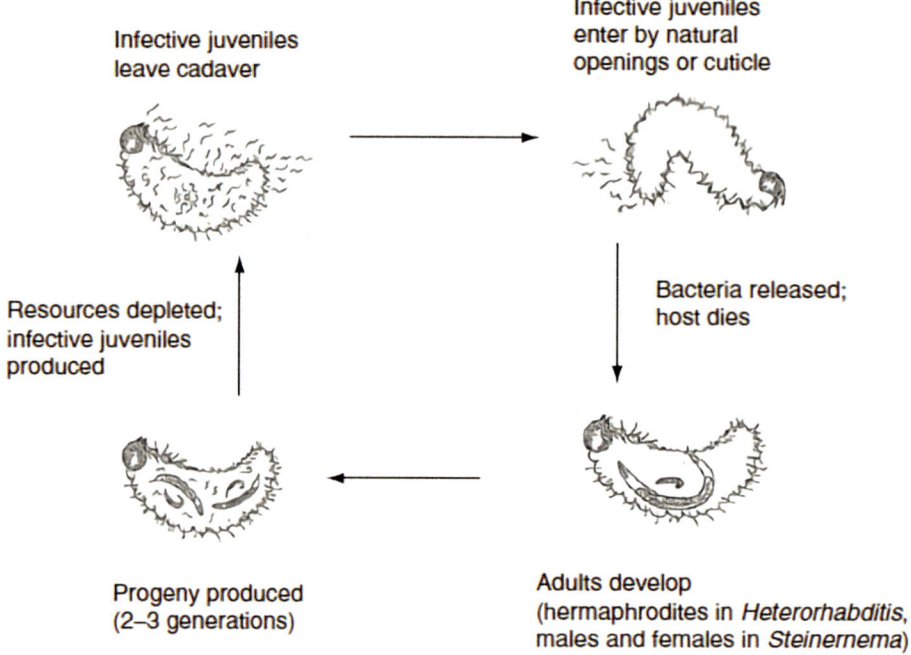

Infective juveniles
leave cadaver

Infective juveniles
enter by natural
openings or cuticle

Resources depleted;
infective juveniles
produced

Bacteria released;
host dies

Progeny produced
(2–3 generations)

Adults develop
(hermaphrodites in *Heterorhabditis*,
males and females in *Steinernema*)

Fig. 4.1 Simplified life cycle of entomopathogenic nematodes (from Griffin *et al.*, 2005).

4.4.8.1 *ACTIVITY: Assess the form of EPN and the viability of an EPN product*

Time required: 1 h
Target audience: Field technicians
Location: Classroom

MATERIALS

- EPN product
- Petri dishes
- Microscope
- Tap water
- Pasteur pipettes/dropper
- Paper for note taking

PROCEDURES

Demonstrate that EPNs are living organisms by applying the following steps:

- Dilute the EPN product with tap water (e.g. final concentration of 1×10^2/ml).
  For example, place 5 g of nematode (or nematode formulation) into 200 ml of water, and then further dilute 10 ml into a total volume of 1 l.
- Leave for 2–3 min.

- Place an aliquot of the EPN solution in a petri dish, then observe under the microscope.
- Participants should write down their observations.
- Ask the participants to comment on what they saw.
  Straight nematodes are dead; hooked and sickle forms might still be living (look for any movement), S-shaped are definitely alive.
- Close the activity by emphasizing that EPNs are living organisms and inform the participants about the conditions under which they perform best.

For a more detailed activity, differentiate between the dead and the living nematodes, and assess the quality of the product provided:

- View under the microscope and count the number of nematodes in a minimum of four 1 ml samples. There should be approximately 50–100 nematodes per sample. If there are more than this, carry out a further dilution as required.
- Count all the nematodes in a sample, and note the number of living [l] and dead [d] individuals.
  Straight nematodes are dead; hooked and sickle forms might still be living (look for any movement), S-shaped are alive.
  Calculate the average number of nematodes from the four subsamples (Øl)
- Do a further four replicates, giving five replicates in total, each with four subsamples.
- Calculation:
  Living nematodes per g of (formulated) nematodes [C]

$$C = \frac{\text{Øl} \times 1000\text{ml} \times 200\text{ml}}{1\text{ml} \times 10\text{ml} \times 5\text{g}}$$

  Øl = Average nematode count
- Number of nematodes per gram should be recorded.
- Also calculate the percentage of living vs dead nematodes.
- As an indicator of quality, the viability of nematodes (percentage alive) should be 80% or more.

4.4.8.2 ACTIVITY: Demonstration of the impact of EPN infection on soil pests

Time required: Set-up 3–4 days, classroom demonstration 1 h
Target audience: Field technicians
Location: Classroom

MATERIALS:

- EPN product
- Tap water

- Filter paper
- Pasteur pipettes/dropper
- Flat containers such as petri dishes
- Soil pests such as cutworms or white grubs
- Microscope
- Pens and paper for note taking

PROCEDURES:

To prepare for the demonstration, follow these steps:

- Dilute the EPN product with tap water (adjust concentration to approximately 1×10^3 nematodes/ml).
- Leave for 2–3 min.
- Place a piece of filter paper in the bottom of a flat container.
- Using a dropper, put the EPN suspension on the filter paper.
  EPN concentrations of **10** or **50** or **100** or **200** or **400** in 1 ml of sterile tap water should be applied to the centre of the filter paper.
- Place 1 to 5 insect larvae into the container (depending on size and availability; field-collected larvae or commercially available hosts can be used as an alternative, e.g. *Tenebrio molitor* or *Galleria melonella*).
- Cover with another piece of filter paper and keep moist (make sure there is no spare water in the dish, as this can increase the risk of bacterial/fungal infection).
- Replicate the above procedure for insect larvae in separate containers but without using EPN, to act as a control.
- Keep all the containers with the insect larvae (both with and without EPN) in a dark and cool place.
- Set up sufficient replicates for distribution to participants (either individually or in groups).
- After 24 h, repeat the above steps with another set of tests and controls.
- The demonstration should take place within 24 h of setting up the last set of petri plates.
- For the demonstration, arrange the containers in groups: control plates, plates with insects that have been exposed to EPN for 24 h and plates with insects that have been exposed to EPN for 48 h.
- Ask the participants to write down what they observe.

Hours after exposure	Condition of the soil pests (e.g. living or dead?)	Colour of the soil pests
Control		
24 h		
48 h		

Note for the facilitator: Pre-test material to determine the actual time it takes for death to occur, as this may vary between nematode species.

In general, the larvae will be dead after 24 h with no colour change. After 48 h they will change colour:

- *Steinernema* spp. - generally grey/brown colour equally over entire body
- *Heterorhabditis* spp. - generally red to dark red-brown colour equally over entire body

This is indicative of the symbiotic bacteria colonization.

Alternative procedure:

To ensure colonization by the EPN it may be applied directly to the insect larvae.

Instead of applying EPN to the filter paper, moisten the filter paper, but add nematodes directly to the insect larvae using a dropper/Pasteur pipette. EPN suspensions with 10 or 20 or 40 or 50 nematodes should be applied in 1 ml sterile water directly on to the insect body.

4.4.8.3   ACTIVITY: Applying EPN in the field

Time required: 1 h
Target audience: Field technicians, farmers
Location: Tobacco field

MATERIALS:

- EPN product
- Petri dishes
- Tap water
- Bucket
- Spray equipment
- Paper for note taking

PROCEDURES:

Begin by emphasizing to the participants that the EPNs are living organisms and explain that the right environmental conditions are important for the survival and success of EPN.

- **Moisture:** Sufficient soil moisture is required for the nematodes' efficacy and survival. EPNs use the water channels like roads to reach their hosts. Therefore, the soil needs to be moist below the level of the grubs. Keeping the soil wet for 2–3 days after an application can help to ensure EPN efficacy.
- **Texture:** Different soil textures have varying capacities for oxygen, which is an important factor in the nematodes' survival. Different soil types can influence the efficacy of the nematodes; the lowest survival for EPNs is recorded in clay soil (compared with sand, sandy loam and clay loam). This lower survival rate is probably due to the lower oxygen levels present in the small pores of clay soils. Note: if the soil is saturated from

heavy rains or lack of drainage, the nematodes could die from lack of oxygen.

- **Temperature:** The effect of temperature on survival varies with nematode species and strains. Soil temperature determines the activity and efficacy of EPNs. If it is too cold, they are inactive and will not actively seek their hosts. Conversely, if the soil is too warm, they will use up their energy source too quickly.
- Check viability of the product:

  - Check product label for expiry date.
  - Confirm product has been transported and stored appropriately before arriving in the field, e.g. in a cool box with icepacks, kept refrigerated.
  - Check product for visual contamination by fungi.
  - Check odour of product; an earthy smell is healthy.
  - If possible conduct a laboratory viability test as indicated in the activity above

Lead the participants through the following steps in order to demonstrate how to apply EPN in the field:

- Irrigate the plot thoroughly (5–8 cm deep) before the application of EPN.
- EPN product (on sponge or inert carrier) should be mixed with water to give required final nematode concentration.
- Stir the nematode solution vigorously to suspend nematodes equally, as they are heavier than water.
- Apply with knapsack sprayer.
- Note to the participants that they may not see the insect cadavers in the field, as they tend to disintegrate in the soil.

4.4.9   Pheromones for monitoring insect pests in tobacco

BACKGROUND

- **Active substance:** Pheromone blend appropriate for the target pests
- **Substance group:** Semiochemicals
- **Target pests:** Pheromones have been commercialized for a wide range of field pests such as *Helicoverpa armigera*, *Heliothis virescens*, *Phthorimaea operculella*, *Spodoptera exigua*, *Spodoptera frugiperda*, *Spodoptera litura* and *Tuta absoluta* as well as stored product pests such as cigarette beetle, tobacco moth and warehouse moth
- **Application site and/or stage:** Seedbed, field, storage
- **Special considerations for use:** Apply in the evening or early in the morning. Do not apply in full sun.
- **Overview of the regulatory status:** Pheromones have been registered in many countries globally, and in some countries, e.g. the USA, they are exempted from the registration requirement.

Pheromones are chemicals produced by animals to attract a mate. For many species of insects, females emit pheromones to attract males. Pheromones tend to be highly specific, so only males of the same species are attracted to the pheromones that the females emit. Synthetic blends of pheromones are available commercially in many countries globally, e.g. Brazil, South Africa and the USA.

Pheromones are used for monitoring insect populations, helping farmers to decide when to apply CPAs, and, for some insect species, can be used for control, through mass trapping or by applying pheromones in the field to confuse the males so that they are unable to locate females. However, mass trapping or mating disruption is generally more likely to be effective if used over a large area. In areas with many small farms, this would be difficult to organize.

Pheromones can be used in the seedbed, field or warehouse to monitor populations of armyworm, tobacco moth, click beetle and cigarette beetle.

Pheromones are low risk, and present no known hazards to humans or the environment. Because they are species-specific and tend to be released in small quantities, no adverse impacts on non-target organisms are expected. The frequency with which the trap should be checked, and the threshold for action should be applied, will depend on the pest in question and the local conditions.

Guidelines for use:

- Use 10 pheromone traps/ha (or double that if infestation persists).
- Replace the lures every 2 months.
- Lures should be stored below 5°C.
- Use a delta or funnel trap, depending on the pest:

  - To mount the delta trap, attach the side flaps to the lower fittings. The glue refill is placed on the inside of the trap with the glue uppermost. Place the pheromone septum on top of the glue in the centre of the refill.
  - The funnel trap is already mounted and the septum should be trapped at the top.

- Check the trap every week and count the insects captured

4.4.10   Matrine for the management of insect pests in tobacco

BACKGROUND

- **Active substance:** Matrine
- **Substance group:** Botanical
- **Target pests:** budworm, aphids, thrips, whiteflies and spider mites
- **Application site and/or stage:** Seedbed, field
- **Special considerations for use:** Apply in the evening or early in the morning. Do not apply in full sun.

- **Overview of the regulatory status:** Matrine is registered for use in to-bacco in some countries in Africa and Asia

Matrine is an extract found in plants from the *Sophora* group of plants, a rela-tive of peas and beans. It can be used to control pests such as budworm, aphids, thrips, whiteflies and spider mites. The use of matrine on tobacco does not have any negative impact on the quality of harvested tobacco.

Matrine acts on the insects' central nervous system, affecting breathing and movement. It is not classed as a highly hazardous pesticide. However, it is harmful if swallowed, and can cause serious eye irritation.

Guidelines for use:

- Matrine should be applied in the seedbed and field at the first sign of the pest, or before if possible, such as in the case of aphids.
- Follow the dosage on the label. For most pests, 60 ml of 0.3% solution in 40–50 l water per acre is effective.
- Spray uniformly over the whole plant. For aphids, ensure that the backs of the leaves are covered.
- Do not spray matrine when rain is expected, or when it is too hot. It is advisable to spray after 4 pm.
- Do not spray near waterways, to avoid the spray affecting fish and other aquatic life.
- Wear protective equipment when handling matrine, including protective eyewear and a mask.

4.4.11   Mexican tea extract for the management of insect pests in tobacco

BACKGROUND

- **Active substance:** Extract of 'American wormseed' or 'Mexican tea' (*Dysphania ambrosioides*, formerly known as *Chenopodium ambrosioides*)
- **Substance group:** Botanical
- **Target pests:** Insect pests including aphids, whiteflies and caterpillars
- **Application site and/or stage:** Seedbed, field, storage
- **Special considerations for use:** Apply in the evening or early in the morning. Do not apply in full sun or when rain is expected.
- **Overview of the regulatory status:** Products containing extracts of *D. ambrosiodes* are registered in several countries in the Americas

The leaves of the flowering plant 'American wormseed' or 'Mexican tea' (*Dysphania ambrosioides*) contain an extract that can be used as a biopesti-cide. It is effective against insect pests, including aphids, whiteflies and cat-erpillars. Extracts of Mexican tea are prepared by farmers in some countries and available commercially in some other countries, like the USA.

The active ingredient in these extracts softens the cuticles of target in-sects which then disrupts the insects' breathing. Mexican tea extract is not classed as a highly hazardous pesticide. However, it is toxic if swallowed or, upon skin contact, can cause skin irritation and allergies, and may be fatal if it enters airways. It is not persistent in the environment, typically degrading within 10 min of application to plants. However, it is very toxic to aquatic organisms, with long-lasting effects.

Guidelines for use:

- Mexican tea extract can be applied in the seedbed or field.
- Begin to apply at the first sign of insect pressure.
- Apply using a pressurized, greenhouse or handheld sprayer.
- Dilute to create a 0.5% solution, or a 1% solution for heavy pest infestations.
- Shake container well before use.
- Spray upper and lower leaves uniformly and to saturation, but not beyond, to avoid excessive runoff.
- Restricted re-entry interval is 4 h.
- Re-apply every 1–2 weeks, as necessary up to a maximum of 10 times per year.
- Do not apply close to waterways, and do not apply when the wind exceeds 10 mph, or when rain is expected within 2 h.
- When applying Mexican tea extract, wear a long-sleeved shirt, long pants, waterproof chemical-resistant gloves and protective eyewear. When applied in a closed setting such as a greenhouse, it will be necessary to wear respirator with an organic vapour (OV) cartridge/canister with an R, P or HE filter.

References

Bateman, M., Chernoh, E., Holmes, K., Grunder, J. and Grossrieder, M. (2016) *Training Guide on Integrated Pest Management in Tobacco*. CAB International, Wallingford, UK.

Copping, L.G. (2009) *A World Compendium: The Manual of Biocontrol Agents* (Fourth Edition). British Crop Production Council, Alton, UK, p. 851.

Ghewande, M.P., Desai, S., Narayan, P. and Ingle, A.P. (1993) Integrated management of foliar diseases of groundnut *(Arachis hypogaea* L.) in India. *International Journal of Pest Management* 39(4), 375–378.

Griffin, C.T., Boemare, N.E. and Lewis, E.E. (2005) Biology and behaviour In: R. Ehlers, P. Grewal, and D. Shapiro-Ilan (eds) *Nematodes as Biocontrol Agents*. CABI Publishing, Wallingford, UK, pp. 47–64.

Lokanadhan, S., Muthukrishnan, P. and Jeyaraman, S. (2012) Neem products and their agricultural applications. *Journal of Biopesticides* 5, 72–76.

CABI – who we are and what we do

This book is published by **CABI**, an international not-for-profit organisation that improves people's lives worldwide by providing information and applying scientific expertise to solve problems in agriculture and the environment.

CABI is also a global publisher producing key scientific publications, including world renowned databases, as well as compendia, books, ebooks and full text electronic resources. We publish content in a wide range of subject areas including: agriculture and crop science / animal and veterinary sciences / ecology and conservation / environmental science / horticulture and plant sciences / human health, food science and nutrition / international development / leisure and tourism.

The profits from CABI's publishing activities enable us to work with farming communities around the world, supporting them as they battle with poor soil, invasive species and pests and diseases, to improve their livelihoods and help provide food for an ever growing population.

CABI is an international intergovernmental organisation, and we gratefully acknowledge the core financial support from our member countries (and lead agencies) including:

UKaid
from the British people

Ministry of Agriculture
People's Republic of China

Australian Government
Australian Centre for
International Agricultural Research

Agriculture and
Agri-Food Canada

Ministry of Foreign Affairs of the
Netherlands

Schweizerische Eidgenossenschaft
Confédération suisse
Confederazione Svizzera
Confederaziun svizra
Swiss Agency for Development
and Cooperation SDC

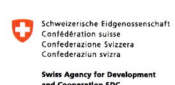

Discover more

To read more about CABI's work, please visit: **www.cabi.org**

Browse our books at: **www.cabi.org/bookshop**,
or explore our online products at: **www.cabi.org/publishing-products**

Interested in writing for CABI? Find our author guidelines here:
www.cabi.org/publishing-products/information-for-authors/